ENDANGERED SPECIES
THREATENED CONVENTION

ENDANGERED SPECIES
THREATENED CONVENTION

THE PAST, PRESENT AND FUTURE OF CITES,
the Convention on International Trade
in Endangered Species of Wild Fauna and Flora

Edited by JON **HUTTON** and BARNABAS **DICKSON**

AFRICA RESOURCES TRUST

EARTHSCAN

Earthscan Publications Ltd, London

First published in the UK in 2000 by
Earthscan Publications Ltd

Copyright © Africa Resources Trust, 2000

All Rights Reserved

A catalogue record for this book is available from the British Library

ISBN: 1 85383 636 2 paperback
      1 85383 667 2 hardback

Typesetting by Composition & Design Services, Minsk, Belarus
Printed and bound by Creative Print and Design (Wales)
Cover design by Susanne Harris

For a full list of publications please contact:

Earthscan Publications Ltd
120 Pentonville Road
London, N1 9JN, UK
Tel: +44 (0)20 7278 0433
Fax: +44 (0)20 7278 1142
Email: earthinfo@earthscan.co.uk
http://www.earthscan.co.uk

Earthscan is an editorially independent subsidiary of Kogan Page Ltd and
publishes in association with WWF-UK and the International Institute for
Environment and Development

This book is printed on elemental chlorine free paper

Contents

Part IV The Future of CITES

Part V Endpiece

List of Acronyms and Abbreviations

CAMPFIRE	Communal Areas Management Programme for Indigenous Resources (Zimbabwe)
CBD	Convention on Biological Diversity
CITES	Convention on International Trade in Endangered Species of Wild Fauna and Flora
COP	Conference of the Parties
DNPWLM	Department of National Parks and Wild Life Management (Zimbabwe)
ESA	Endangered Species Act (US)
EU	European Union
FFI	Fauna & Flora International
GATT	General Agreement on Tariffs and Trade
GEF	Global Environment Facility
HSUS	Humane Society of the United States
IIED	International Institute for Environment and Development
ITRG	Ivory Trade Review Group
IUCN	The World Conservation Union (formerly the International Union for Conservation of Nature and Natural Resources)
MQS	management quota system
NGO	non-governmental organization
Resolution Conf	Resolution passed at a meeting of the Conference of the Parties to CITES. (Each resolution is identified by two numbers separated by a period. The number before the period indicates which meeting the resolution was passed at, the number after the period indicates which resolution it was. So Resolution Conf 9.24 is the 24th resolution passed at the Ninth Conference of the Parties.)
SACIM	Southern African Centre for Ivory Marketing
SACWM	Southern African Convention on Wildlife Management

SADC	Southern African Development Community
SADCC	Southern African Development Coordination Conference
SBSTTA	Subsidiary Body on Scientific, Technical and Technological Advice (Convention on Biological Diversity)
SSC	Species Survival Commission
TRAFFIC	The wildlife trade monitoring programme of WWF and IUCN
UCDA	ursodeoxycholic acid
UNCED	United Nations Conference on Environment and Development
UNEP	United Nations Environment Programme
WTMU	World Trade Monitoring Unit
WTO	World Trade Organization
WWF	World Wide Fund For Nature
ZSL	Zimababwe Sun Limited

About the Contributors

Morné A du Plessis was assistant director of Biodiversity at the Natal Parks Board, South Africa, during the mid-1990s. In late 1996 he was appointed director of the Percy FitzPatrick Institute at the University of Cape Town which, in addition to its strong ornithological reputation, offers a highly acclaimed coursework Masters degree in Conservation Biology.

Barnabas Dickson has taught philosophy at universities in Britain and Zimbabwe. He specialises in environmental philosophy and currently works as an environmental consultant.

Jon Hutton, director of the Africa Resources Trust, has over 20 years of experience in wildlife management in southern Africa. He was introduced to the complexities of CITES in the late 1980s while helping develop crocodile management regimes in a range of African countries on behalf of the CITES Secretariat. From 1992 until 1997 he represented Africa on the CITES Animals Committee and was a member of the Zimbabwean delegation at three Conferences of the Parties.

Chris Huxley is a zoologist who has been deeply involved in CITES work for over 20 years. This started in 1978 when he went to Hong Kong to set up and run the Hong Kong Government CITES Management Authority. Since then he has worked in the CITES arena for the CITES Secretariat, TRAFFIC, WTMU, SADCC/SADC, SACIM, Cathay Pacific Airways, IUCN, WWF and FFI.

Robert Jenkins presently occupies the chairmanship of the CITES Animals Committee – a position that he has held since 1992. He resides in Canberra, where he is employed as director of International Wildlife Unit with Environment Australia. He is also formally associated with implementing the IUCN Sustainable Use Initiative in the Asia–Pacific Region and has collaborated with numerous governments throughout the world to develop management systems for the sustainable use of wild fauna.

Henriette Kievit has formerly worked for the Africa Resources Trust. She is now a business consultant for the Amsterdam Consulting Group. She specialises in trade relations between southern Africa and the EU.

Rowan Martin served in the Zimbabwe wildlife department for 25 years, rising to deputy director (Research). He carried out two major consultancies for CITES (one on the ivory trade in 1985 and one on leopards in 1987). He represented Zimbabwe at CITES meetings and on the CITES Standing Committee. He was responsible for numerous resolutions adopted by CITES, including one on the benefits of trade and another on new criteria for listing species on the CITES appendices.

Simon Metcalfe has worked in the field of rural development in southern Africa for the past twenty years. Working for the Zimbabwe Trust he helped initiate CAMPFIRE in Zimbabwe, which empowered communities living near protected areas with rights over the wildlife on their land. He is presently a freelance consultant working on community-based natural resource management issues in southern Africa.

Phyllis Mofson is a senior analyst at the Florida Legislative Committee on Intergovernmental Relations. She has previously held posts at the United States Department of State and the Florida Department of Community Affairs. Her doctoral dissertation examined the relationship between CITES as an international regime and two of its member states, Zimbabwe and Japan.

Marshall Murphree was born in Zimbabwe and educated at universities in the United States and Great Britain. He was Director of the Centre for Applied Social Sciences at the University of Zimbabwe from 1970 to 1996 and is now Emeritus Professor of Applied Social Science. He was Chairman of Zimbabwe's Parks and Wildlife Board from 1992 to 1996.

Timothy Swanson is professor of Environmental Policy, University College London, and author of several books and many articles on the regulation of endangered species, including *The International*

Regulation of Extinction (1993), *Global Action for Biodiversity* (1997) and *A Global Framework for Biodiversity Conservation* (1998).

Michael 't Sas-Rolfes is a conservation economist, based in Cape Town, South Africa. His work includes research on wildlife trade issues and innovative ways of ensuring the economic viability and sustainability of protected areas. He has written numerous papers and articles, including *Rhinos: Conservation, Economics and Trade-Offs* (Institute of Economic Affairs, London) and *Who Will Save the Wild Tiger* (Political Economy Research Center, Bozeman, Montana).

Grahame Webb is an Australian zoologist with 25 years' experience on crocodile conservation, management and research in Australia and other countries. As director of Wildlife Management International Pty Limited, he has been involved in extending the sustainable use principles involved with crocodile conservation to other species. He has worked on a variety of CITES issues, ranging from the swiftlets of south-east Asia that are used for bird nest soup through to the sustainable harvest of sea turtles in Cuba. He has published widely on crocodiles and the sustainable use of wildlife generally, and he is vice-chairman of the IUCN Crocodile Specialist Group (Eastern Asia, Oceania, Australasia) and chairman of the IUCN Australia New Zealand Sustainable Use Specialist Group.

Acknowledgements

The editors wish to thank the trustees, management and staff of both the Africa Resources Trust and the Zimbabwe Trust, without whose enthusiastic support this book would not have been possible. We hope that, in return, we have made a positive contribution to their public awareness campaign that highlights the important relationship between international trade and the rights and aspirations of rural communities seeking to improve the quality of their lives through the sustainable use of natural resources. The editors are also grateful for the support and facilities provided by the Department of Geography at the University of Cambridge.

Introduction

Jon Hutton and Barnabas Dickson

The Convention on International Trade in Endangered Species of Wild Fauna and Flora (CITES) was originally signed in 1973 and came into force two years later. The aim of the convention was to save wild species from extinction and the means adopted to achieve this were the regulation and restriction of the international trade in wildlife. Its policies are decided on at the Conference of the Parties (COP) which meets roughly every two years. For most of its existence, CITES has been the major tool possessed by the international community for preventing the loss of species, and high expectations have been placed on it.

CITES has proved to be the most controversial of the international environmental conventions. At successive meetings debate has raged over the most basic assumptions of the convention. Some have explained the apparent failures of the treaty in terms of a failure to effectively enforce its regulations, while others have seen the CITES approach to conservation as fundamentally misconceived. Since much of the remaining biodiversity is located in countries of the South, and concern about the loss of diversity is most acute in the North, there has been ample scope for North–South conflict.

One-quarter of a century after the convention was first signed, the time is ripe for an assessment of its successes and failures. Have the aims of CITES been achieved? Are there any alternative approaches to wildlife conservation that would be preferable? In what direction should CITES evolve? For all the public and media attention attracted by CITES there has been surprisingly little critical analysis of the convention published in an accessible form. This book is intended to help fill that gap. From this study, four developments over the last 25 years emerge as being of crucial importance.

Most centrally, there have been improvements in our understanding of the threats to wild species. The convention is founded on the assumption that the international trade in wildlife is an important threat to their continued existence. Indeed, it is the only threat it addresses. The prescription that CITES offers is a mixture of bans on international trade for the most endangered species, and regulation of the trade in less seriously threatened species. Part of the weakness of CITES is that it has not always been successful in enforcing its bans and regulations. Where it has attempted to ban trade, illegal trade has often flourished, and where trade has been allowed, CITES has often been unable to regulate it effectively. A much more serious difficulty is that for many species international trade is not the primary threat. It has gradually been recognized that other processes, in particular the loss of suitable habitat, is much more significant. Thus, for numerous species, the CITES remedy will be quite inappropriate. Indeed, there is an argument that the policies offered by the convention have actually exacerbated the problem. By restricting trade in wild species, and so limiting the benefits that humans can derive from them, CITES has actually reduced the incentive to maintain wildlife habitat. This has hastened the decline of species.

The growing acknowledgement of the importance of habitat loss as a threat to wildlife has led some to the conclusion that the human use of wildlife, and commercial trade in particular, can actually be a positive force for conservation. If people can benefit from wildlife, they have an incentive to maintain wild habitat and not to convert it to other uses such as agriculture. According to this school of thought, the key issue is whether the off-take of a species is sustainable in the long-term, not the use to which the species is being put. This challenges the basic assumptions contained in CITES, and has also provoked the anger of those who would oppose the use of wildlife on any grounds, even if it serves conservationist ends. For this constituency, the commercial exploitation of nature is simply an inappropriate way to relate to it. The debate about the effectiveness of CITES has quickened in recent years and the convention has gone some way towards recognising, albeit in a piecemeal way, the conservation benefits that use can have. Questions remain, however, about how far the convention, as currently structured, can change.

A second development has been that countries of the South have become more forceful in putting forward their own case. The original convention was largely developed by conservationists from the North and reflected their conception of the problem. Southern African countries have been particularly prominent in promoting a new perspective on conservation. Many wish to distance themselves from the preservationist approach that they see as a legacy of the colonial period and they emphasise that if conservation is to be successful it must provide tangible benefits to those who live closest to the wildlife. This view has been criticised by those who see it as providing a licence for the unregulated exploitation of wildlife.

The third development has been an increasing emphasis on the social dimension of conservation. As the limitations of the assumption that trade is the chief threat to species have become apparent, so the need for a more sophisticated understanding of the complexity of the interactions between wildlife and human society has become more pressing. This has drawn social scientists into the debate about the future of CITES and conservation. Another aspect of the social dimension is the growing recognition that, because the fate of wildlife is so closely entwined with changes in human society, a policy for wildlife is simultaneously a policy for human society, raising questions of justice and equity within our own society. There was no acknowledgement of this in the original convention and there has been some reluctance among conservationists to engage with these evaluative, 'non-scientific' issues. But the growing popularity of the notion of sustainable development, with its acknowledgement of a linkage between environmental and social concerns, has exerted an influence on the debate about wildlife.

A final development was the signing of the Convention on Biological Diversity (CBD) at the United Nations Conference on Environment and Development (UNCED) in 1992. As with CITES, the CBD is concerned with the loss of species, but it is a more comprehensive convention and one which takes into account the lessons of recent years. It does not focus on just one threat to wildlife, and it does not offer just one remedy. This opens up the issue of the institutional future of CITES.

The above four developments are closely linked. The case for acknowledging the multiplicity of threats to species and recognizing the importance of sustainable use has been made forcefully by countries from the South. They have also been most ready to connect issues of conservation with those of equity. The signing of the CBD was, in turn, a symptom of these first three developments. Together, these four developments have placed the future of CITES in question.

While most of the contributors to this book would recognize the developments as crucial ones, they are far from being unanimous on all the issues. The differences become most pronounced when the implications of these developments for the future of the convention are considered. Some would see a large role for a suitably reformed

CITES in ensuring that all trade is sustainable, while others would put a much greater emphasis on institutional changes at the local level. Even among those who would emphasize the concepts of human centred conservation and sustainable use, there is room for significant debate about how it should be realized.

The volume is organized around four main sections. The first examines some of the essential background issues. Huxley contends that the original convention was motivated by a concern to control the illegal trade in wildlife. He argues that its ability to achieve this aim was hampered first by a lack of understanding of how to operate the necessary controls and, later, by attempts to change the direction of the treaty. Then, du Plessis discusses the fundamental issue of where the threats to species are coming from. He identifies habitat loss as posing the biggest threat, with international trade being a threat in a much smaller number of cases. He points out that this conflicts with the presumption of the convention that trade is the most significant threat.

The second section looks at how CITES has operated in practice. Martin points out that on a direct assessment of species that have been saved, the record of CITES is unimpressive, although the evidence is inconclusive. If one steps back and identifies the conditions under which an international conservation convention is likely to succeed, it appears that CITES is not well designed. The other chapters in this section tackle more specific questions. Dickson identifies the different versions of the precautionary principle that are embodied in CITES and assesses what role the principle might legitimately have within the treaty. He argues that appeal to the precautionary principle alone cannot be used to justify any particular policy response. Jenkins focuses on the problems associated with Appendix II of the convention. The treaty stipulates that an Appendix II species can only be traded when the trade will not be detrimental to the species. In practice this condition has frequently been violated, with the consequence that Appendix II species frequently become more endangered. Jenkins suggests that, with the emergence of the significant trade process, CITES has begun to tackle this problem and is moving towards an implicit recognition of the positive role played by use. Jenkins' chapter illustrates how attempts to address the weaknesses of CITES can sometimes lead in the direction of fundamental reform. Hutton tackles the problematic question of stricter domestic measures. Enshrined in the treaty is a right for any Party to the convention to take stricter conservation measures than those agreed to by all the other Parties. Hutton argues that economically powerful countries have appealed to this right to justify unilateral measures based on misguided conceptions of where the threat to species comes from. He recommends a move towards multilateral measures.

The third section consists of case studies, mainly of how CITES has dealt with particular species. First, 't Sas-Rolfes compares CITES record on four different animals. His conclusion is that in most cases, CITES has not worked well. One of the animals 't Sas-Rolfes discusses is the elephant, which has absorbed a huge amount of CITES time and resources. In consequence, discussion about the convention has also tended to concentrate on elephants. But other species also offer useful and different lessons. Kievit examines the Nile crocodile. She charts the way in which CITES, despite initial resistance, has slowly moved towards a pro-use position. In the following chapter Webb compares the way the convention has treated the non-charismatic crocodiles with a very different attitude to the much more charismatic sea turtles. For the latter group there has been extreme reluctantance to sanction any form of use at all, despite the similar biology and conservation needs of the two groups. Webb attributes this to the charismatic appeal of sea turtles, rather than to any scientifically based considerations. In a different sort of case study, Mofson examines the politics of CITES. She focuses on the way in which Zimbabwe, an economically weak state, has come to play a key role in shifting the parameters of debate within CITES, towards a pro-use position.

The fourth section looks at the future of CITES and wildlife conservation, in the light of the developments that have been charted earlier in the book. All of the authors in this section recognize the importance of sustainable use but differ, at least in emphasis, in how this is to be guaranteed. Martin discusses the institutional relationship between CITES and the CBD. He puts the case for subsuming the older treaty under the newer, more comprehensive one. Swanson argues that the sustainability of use can be guaranteed through a constructive trade control regime. He suggests that a suitably reformed CITES could play an important role in instituting and maintaining such a regime. In contrast, Metcalfe sees the key lying in the devolution of control over wildlife to local people and their consequent empowerment. He appears to accord a much lesser role to CITES, but suggests that market liberalization poses a significant threat to the success of communal management schemes. Dickson offers a comparative assessment of the merits of the global regulation and communal management approaches to wildlife conservation. He sets this in the context of the historical evolution of conservation since the colonial era.

In the final section, Murphree reflects on the implications of different models of sustainable use for one particular rural community in Zimbabwe. In his view the emphasis should be placed on community management, but he maintains that this requires a supportive regulatory environment.

Part I

BACKGROUND

Chapter 1

CITES: The Vision

Chris Huxley

INTRODUCTION

CITES, known in its early days as the 'Washington Convention', has become the world's best known wildlife conservation convention. Since it was originally signed in 1973 it has received a great deal of public attention and media interest for a combination of reasons.

First and foremost, CITES deals with highly emotive issues such as: the exploitation of nature for profit; the trafficking in illegal goods; the killing or capture of wild animals; and the use of these animals for what some would regard as abhorrent purposes. Secondly, it is one of the few international vehicles for wildlife conservation that at least appears to provide some much-needed action. The inclusion of particular species in the appendices of CITES has often been heralded as a triumph for conservation, and the imposition of trade bans and the seizure of shipments of illegal specimens have both been seen as positive contributions to international conservation. The fact that none of these measures is real conservation action is another matter – it is the perception that something is being done that has led to CITES' unique position. Finally, it has been recognized by many professionals working in the field of wildlife conservation as a remarkably potent tool which, if used well and applied correctly, could lead to substantial progress in halting the over-exploitation of wildlife resources.

The huge interest in the convention was initially generated mainly by western societies or western cultural values. This has had a profound effect on the way in which it has developed, as well as promoting the false impression that actions of the treaty are real conservation actions. It is this latter point that has given rise to rifts between the various groups that have been involved in CITES and to a frequent failure to recognize that CITES is not an end in itself, but a tool to be used to assist real conservation actions. So, how did CITES come to occupy this pre-eminent position? How did it come into being? What was it originally intended to achieve and how? These are questions that need to be addressed if we are to understand CITES and how it has developed into the treaty that many people know of, but which very few really understand.

BACKGROUND

Man's exploitation of wildlife for profit is not a recent phenomenon and the international trade in wildlife has been widespread for many centuries. In earlier times, some of this trade caused the decline of wildlife populations, but this was neither as frequent as in this century, nor regarded as a matter of any great concern. The world was a bigger place then, with fewer people. The critical factors that changed were the huge expansion in the human population (both in numbers and in geographical space) and the rapid development of modern systems of transport and communications. Together, these have acted to increase the rate and scope of man's exploitation of wildlife, as well as allowing information about it to become more widely available.

The very rapid increase in man's ability to exploit wildlife commercially for international trade, coupled with a concomitant increase in demand for wildlife and its products in the developed world, led to what seemed to be a series of significant species depletions. These reductions were sometimes localized and at other times widespread. They generally, however, shared the characteristic that they involved species that were easily recognized and 'popular' in the sense that the public (at least in the western world) acknowledged them as being of some aesthetic value. Subsequently, it became apparent that some of these declines had not been as severe as originally portrayed and this was to lead to considerable controversy.

Despite popular belief to the contrary, there have been remarkably few, if any, species extinctions that can be attributed to exploitation for international trade. There is a possible explanation for this. Although commercial exploitation may result in a reduction in numbers, there will come a point, before biological extinction occurs, when the numbers are so reduced, and individuals of the species correspondingly so difficult and costly to locate, that exploitation is no longer commercially viable. At that point, commercial exploitation will cease. However, for at least some species this explanation will not be valid. In some specialized markets (rare parrots, for example) the demand, and thus the commercial value, will increase with rarity. In these cases commercial exploitation may still be viable even when the numbers become very low, and exploitation may continue until extinction occurs. Nonetheless, it is an observable fact that, at most, very few species have been entirely exterminated as a result of international trade. Many have become extinct as a direct result of exploitation, but this has been for purposes other than international trade. So international trade does not seem to be an important factor in species loss. Despite this, much of the concern behind the creation of CITES was that species would indeed become extinct as a result of this factor. An additional concern was the issue of species depletion (ie the significant reduction of the population of a species) although the significance of this element was apparently not great in the debate that led up to the creation of CITES.

The expression of international concern took a long time to develop any real momentum. In the early part of this century it was a rare occurrence for opposition to be voiced with respect to the exploitation of wildlife for international trade. This was partly because it had not yet become a major and highly visible problem and partly because the wildlife conservation ethic was only just beginning to emerge as a force in international affairs.

THE FIRST STEPS

Although concerted local and national efforts to promote wildlife conservation (many of which were motivated by selfish interests) date back many centuries, it is only relatively recently that such efforts have been made in the international arena. An early attempt to use international legislation to promote wildlife conservation was the 1911 Fur Seal Convention, designed to deal with the problem of over-exploitation of the fur seals of the Pribilof Islands. Further such moves have led to several other conventions, including the International Convention on the Regulation of Whaling in 1946. These, and others, emerged from an increasing recognition that some important wildlife populations were becoming drastically reduced through uncontrolled exploitation for various purposes. This lack of control was clearly a central element motivating the concern.

During the 1950s, conservationists began to worry that the escalating international trade in both live animals and their products constituted a severe threat to species survival. At first this attention focused on a relatively narrow range of species, including the spotted cats (traded for their furs), primates (used in medical research) and crocodiles (killed for their skins). In time the concern widened. By 1960 there was sufficient international impetus for the International Union for Conservation of Nature (IUCN), at its General Assembly, to draw attention to the problem and urge governments to take action to prevent international trade in wildlife taken illegally. It is significant that the first calls for action demanded the establishment of import controls to prevent illegal trade, where illegal trade was understood as trade in wildlife taken in or exported from the country of origin in contravention of that country's laws. Nevertheless, the knowledge of what was actually occurring was very superficial. For example, an early emphasis on trophy hunting as a threat to species developed and it remained a significant element in the events leading up to the establishment of CITES, despite the fact that evidence that this was indeed a threat was scanty.

The IUCN's next General Assembly, held in Nairobi in 1963, passed a resolution calling for an international convention to address the issue. This was followed with the preparation of a first draft of such a convention in 1964. Thereafter, there was substantial consultation involving international organizations such as the United Nations and the General Agreement on Tariffs and Trade. Concern was not limited to wildlife conservationists, as many others interested in the issue for other reasons were also worried. For example, the International Fur Trade Federation imposed a voluntary ban on its members with respect to the trade in certain spotted cat skins and other furs. At the same time, pressure was mounting for governments to take effective action at the national level. Many countries adopted legislation to improve their control of the wildlife trade, with some establishing partial or even total prohibitions. Primates, spotted cats, tigers, otters, rhinos, elephants, crocodiles, turtles and many others were identified as being potentially at risk as a result of unsustainable trade levels. But it remained the case that there were remarkably few data to show what effect exploitation was having in the wild.

A significant development took place in the US with moves to pass domestic legislation to control imports of wildlife. In 1968, this legislation, in the form of the Dingell Bill, failed to pass. By 1969, however, the pressure was such that its successor, the Endangered Species Conservation Act, became law. This Act sought:

> '*to protect the threatened species of the world by banning all imports of such species* whether or not *they can be taken legally in their country of origin*' (*Oryx*, 1970).

This approach was to have a profound influence on the development of CITES and marked a fundamental change in what was being attempted. It was the beginning of the move towards the imposition of trade controls that were based on the importing country's views as to what should be allowed. Little or no account was to be taken of the views of the exporting country. For the US, illegal trade was no longer the sole target. The US government would decide which species were threatened. The major source used to determine what to control (or prohibit) was the IUCN Red Data Books.

The IUCN General Assembly in 1969 and the UN Conference on the Human Environment in Stockholm in 1972 provided the final impetus for the convention. In March 1973, representatives from over 80 states met at a plenipotentiary conference to agree on the final text. The text had been subject to revisions over a period of years and it represented the consensus, among those taking part, of what was needed and how CITES should operate. It was signed by 21 states. At the same meeting around 1,100 species were placed on the appendices. The appendices reflected fairly closely the species listed in the IUCN Red Data Books. By July 1975, the ten state ratifications needed for CITES to enter into force had been deposited and CITES became a reality.

The Birth of CITES

The main thrust of the convention was the establishment of a set of import, export and re-export controls on the species listed in the three appendices. Although other factors are included in the text, it is abundantly clear that the principal aim was to create an internationally accepted system for controlling the wildlife trade in such a way as to reduce, as far as possible, the chance of illegal trade occurring. The preamble of the convention states that:

> 'international cooperation is essential for the protection of certain species of wild fauna and flora against over-exploitation through international trade'.

As this indicates, CITES was established primarily as a mechanism for *international cooperation*. However, while there was general agreement both on the underlying goal (to prevent international trade from causing species extinctions) and on the mechanism to achieve this (international cooperation in trade controls) there was a dearth of information as to the scale of the problem and a lack of understanding of how to operate trade controls. As noted in *Oryx*, 'when CITES was originally set up and signed, few administrators realized that more than a handful of endangered species were involved' (*Oryx*, 1977).

Moreover, few countries had any real experience of how to implement controls of the type required by CITES. Although the US and UK had both been enforcing domestic legislation with similar intent, and many of the exporting countries had laws controlling the export of at least some wildlife, CITES was helping to break new ground in attempting to establish an international mechanism for cooperation in wildlife trade controls. The conservationists who led the initiative knew well enough what the problem was, but they had very little expertise in dealing with it. Of those involved in defining the way in which the convention would operate, there were very few who had any experience at all of designing or implementing trade control measures. Surprisingly, instead of utilizing the existing, well-tried and successful customs network, CITES created an entirely new system based on the designation of 'Management Authorities', which were, most frequently, wildlife conservation agencies. But these Management Authorities often had little or no practical experience or expertise in such work.

Another issue concerned the problem of what products were to be covered by the trade controls. This led to the use of the term 'readily recognizable parts and derivatives'. The use of this vague expression illustrates the lack of any substantial technical basis for what was being attempted. The issue of which wildlife products should be controlled was too difficult to resolve, so it was disguised with a form of wording that allowed each country, at least at first, to include whatever it wished in the controls. A further factor that was to create difficulties at a later stage was the absence of many important countries at the Washington meeting and the representation of many others (particularly those from the developing world) by their diplomatic missions. These missions often had little or no knowledge of, or effective briefing on, the technical issues involved.

The lack of understanding of how to operate a system of trade controls at the inception of CITES left the convention grappling with the issue for many years afterwards. In addition, more fundamental differences were to surface. While the text of the convention implies that illegal trade was the problem and the prevention of illegal trade the solution, others held that the main problem was over-exploitation as a result of increasing demand and that the solution was to eliminate demand and stop the trade, or at least to reduce it. The approach taken by some, especially in the US, suggested that they also believed that legal exploitation in some developing countries should be curbed.

THE FIRST TEN YEARS

During the early years it was quickly realized that the convention required a great deal of interpretation and refinement if it was to work effectively. The debates at the first few meetings showed that a mechanism had been established that was neither clearly defined nor fully understood. One central issue was the criteria that were to be used for including species in the appendices. The original plenipotentiary conference in Washington had not established any criteria. It was only at the first Conference of the Parties in 1976 that explicit criteria were formulated and even these were subsequently regarded as unsatisfactory by many. Some countries found that what they had signed up to was not quite what they had envisaged. Perhaps more importantly, many countries appeared not to take it at all seriously. This was especially true of some of the major importing countries.

The first ten years were spent debating how the convention should operate and which species should be included in the appendices. At first this seemed to be progressing reasonably well, but at the Fourth Conference of the Parties, in Botswana in 1983, the first real stirrings of discontent became apparent. This discontent was typified by Mozambique's consternation to find that, in order to trade in crocodile skins taken legally from an abundant population, they would have to spend a great deal of money proving that a species which was considered a pest was indeed not 'endangered'. Furthermore, the response to technical difficulties in trade control was usually to define more complex and difficult regulations. Thus, the convention began to become very complicated and expensive to operate in many countries. This was of great concern both to developing countries with inadequate resources for implementing complex regulations and those few developed countries that were making genuine efforts to implement the convention in the way originally intended.

An interesting paradox appeared during this time. Although the existence of trophy hunting had been a major factor in persuading people that a problem existed, it was beginning to be recognized that sport hunting could be a positive force for conservation, as the revenues generated from trophy hunting could provide both the means and the incentive to invest in conservation. In this way, devoting land to wildlife could become an economically viable land-use option.

Consequently, one of the elements that had originally given rise to the establishment of CITES began to be seen as an acceptable, perhaps even beneficial, activity.

THE SECOND DECADE

It took ten years of operation of CITES before the major controversies surfaced in any substantial manner. However, once the majority of the world's nations had joined and the basic direction in which CITES was moving became apparent, expressions of unease were voiced.

One major factor was the increasing tendency for the addition of a species to the appendices, especially to Appendix I, to be regarded as, in itself, a positive conservation measure. Western non-governmental organizations (NGOs) found they could use the convention as a fund-raising tool. Crises could be announced, campaigns embarked on and large sums of money raised to 'save' species. The whole process culminated in the successful 'saving' of the species through its inclusion in Appendix I. A further complicating factor was that the main concern of some of these NGOs was with the issues of animal rights and animal welfare, rather than with the conservation of wild species. They viewed CITES as a mechanism for promoting their own philosophy. A large number of those involved in primarily conservation-oriented activities saw this as an attempt to derail or divert the convention from its original direction.

The convention appeared to be moving closer to the view that all trade was dangerous and should be stopped. This was questioned by some. Was it really true that the wildlife trade was in principle a destructive force? Was the knee-jerk reaction of placing every species that was subject to international trade and classified as possibly threatened in Appendix I really justified? Why was it that some of the most visible species for which this had been tried had continued to decline? Surely, the original intention had been to create a mechanism to prevent illegal trade by establishing an international trade control system? Alternative arguments began to appear. They focused on the need for real, concrete conservation action and suggested that what was most important was that there should be sound management on the ground. If effective management included commercial exploitation for international trade, surely it should not be disallowed? This argument was accepted in some cases. Commercial export quotas were agreed for a number of Appendix I species and crocodile ranching became a highly successful conservation activity based on commercial, international trade. But the dissatisfaction continues, because there remains a division between those who want CITES to be a tool of international cooperation to combat illegal trade, and those for whom CITES is more than a tool – perhaps even an end in itself.

CONCLUSIONS

CITES is regarded by many as the world's leading, most successful international conservation convention. It was created as a result of widespread international concern expressed by wildlife conservationists in the 1950s and 1960s. Their worries were that the rapidly increasing demand for wildlife in the developed world was causing severe over-exploitation of some species, mainly in developing countries, and that this would lead to extinctions. Although the problem was recognized in general terms, there was not a great deal of hard information on which to base a strategy for controlling this phenomenon. Nonetheless, there was a consensus that an international convention was needed in order to establish a mechanism for international cooperation in law enforcement. The primary intention of CITES was clearly exactly this, to set up a system through which the trade controls in importing countries could be matched with those in the exporting countries. This had never been tried before to any significant extent and the convention was breaking new ground.

Has CITES achieved this principal aim? To a large extent it has. A global system for controlling wildlife trade is in place and much illegal trade has been eliminated. Does this mean that CITES has been successful in realizing its underlying goal of saving species? Not to any great extent, since although the mechanism for controlling trade is there, over-exploitation of wildlife is still a widespread and common phenomenon in many developing countries. The conclusion we must draw from this is that either CITES was misdirected in the first place, or the application of CITES has been misguided.

Of these two possibilities, the latter is the more likely. CITES has too often been used in a negative manner, rather than as a positive, constructive tool. Instead of using CITES' controls as a mechanism to encourage the sustainable use of wildlife, it has more frequently been the vehicle for prohibition, though this has not been the case in all instances. There are a few examples of how CITES has been truly successful in bringing the wildlife trade under control and then taking that positive, extra step of encouraging legal trade in sustainably exploited species. More typically, CITES has been used in a manner for which it was clearly not intended: to impose or force inappropriate solutions on wildlife conservation problems, especially in developing countries. This ignores the fact that if the protection and management efforts in the field are ineffective, CITES can do little more than monitor the demise of wildlife populations. It is necessary for CITES to facilitate positive action to improve the real protection and management of wildlife.

REFERENCES

Oryx (1970) 'Notes and News', Vol X, No 4, pp 210–11
Oryx (1970) 'Notes and News', Vol XIV, No 2, pp 97–8

Chapter 2

CITES and the Causes of Extinction

Morné A du Plessis

INTRODUCTION

As concerned citizens and biologists we are anxious to understand how natural diversity can be maintained in a world of rapidly diminishing resources – resources that are important to the livelihood of human beings as well as to the millions of other organisms that share this planet. Human activities are causing major changes to the Earth's biota. Extinction, the ultimate change, is occurring today across a broad range of terrestrial and aquatic habitats. Although much of this biodiversity crisis is almost certainly due to human impact during recent centuries, we still do not effectively prioritize conservation action on the basis of what we know about the causes of extinction. In this chapter I shall attempt to show, in particular, that CITES does not focus on the most important threats to wildlife.

EXTINCTION AND PHYLOGENY

Life on this planet has existed for at least 3.5 billion years, a history marked by a series of diversification events. Global biological diversity is probably now at an all-time high. Plant diversity, for example, has been increasing appreciably since about 700 million years ago and increasing rapidly since about 100 million years ago. Such proliferation is evident despite a series of mass extinction events. A comparison of these trends against estimates for the longevity of species is illuminating. Although such estimates are highly tentative, particularly for terrestrial organisms, a range of longevity between 1 and 10 million years for most species seems broadly acceptable. This marked turnover in species signifies that an enormous amount of taxa has become extinct. Indeed, there is general agreement that more than 99 per cent of all the species that ever existed have become extinct. Species extinction is therefore a natural process that occurs without the intervention of man. However, there is little doubt that man has been the cause, either directly or indirectly, of a large number of extinctions. This elevated, man-induced extinction rate far exceeds the background extinction rate and, at present, it appears to be increasing still further.

RARITY AND EXTINCTION

The role of rarity in the recognition and classification of threatened and endangered species is unclear. To some, rarity and the threat of extinction are almost synonymous and it is widely assumed that rarity is associated with increased extinction risk, while others recognize that rarity per se is insufficient to classify a species as being at high risk of extinction. Some rare species persist and they do so because they have biological characteristics that allow them to become rare and yet then enable them to endure in such a state. But where there are processes causing a continual decline in numbers, a species may well be at risk. A better understanding of the causes and consequences of rarity would assist greatly in the setting of conservation priorities.

VARIATIONS ON THE EXTINCTION THEME

Although extinction is usually defined as the total disappearance of a species from the face of the earth and an absence of sightings for a period of at least five decades, it is important to recognize that there are variations on this somewhat strict definition.

Commercial extinction

Commercially important species of mammals, in particular, are susceptible to over-harvesting, sometimes to the point of near extinction. Examples include the blue whale, the right whale, the northern elephant seal, the American bison and the black and white rhinoceros. However, the commercial exploitation of wildlife can lead to two very different end results. For resources that are harvested in large quantities, a point can be reached at which it becomes economically unviable to continue harvesting; this happens with numerous fish stocks. In these instances, commercial unviability can result in a positive feedback loop leading to persistence and sometimes recovery. In other cases, the increasing rarity of an organism may add the prized element of exclusivity to the value of the commodity and this can drive the price even higher, justifying more expenditure on harvesting. This has been the case with rhino horn.

Population extinction

Population extinctions can have important consequences for the species as a whole, especially where marked genetic variation exists between populations or where populations form part of a larger metapopulation. Moreover, species extinction is simply the endpoint of a process of population extinctions. So, if we want to expand our focus from the preservation of endangered species to include the prevention of endangerment in the first place, we should study population extinctions as well. In addition to being an important step on the path to species extinction, population extinctions can have important consequences of their own. The loss of local populations means the loss of the functional role of these species. This may be particularly important where the species are keystones or 'ecosystem engineers'. These are the species that play an especially important role in shaping and maintaining an ecosystem. Extinctions of populations of economically important species can also have direct economic and social effects.

Regional extinction

Species conservation activities are usually effected and coordinated at the level of individual states. So, extinction from these politically defined regions may attract particular attention from conservation agencies within those boundaries. But it is also the case that the response by nation-states may not reflect the seriousness of the threat to the species as a whole. For example, even though less than 10 per cent of the Blue Swallow's (*Hirundo atrocaerulea*) distribution range (and total population) formerly fell within South African borders, its decline has attracted disproportionately large conservation resources relative to other declining species in that country.

THREATS

Instead of concentrating on global extinctions (which are often difficult to establish unequivocally) it is vital to assess and monitor the status of, and threats to, both species and their habitats if global trends of species diversity are to be established. To this end, IUCN has developed a quantifiable set of criteria to assist with the task of assigning an objective threat status to species of concern (Mace and Collar, 1994). Thinking in terms of the threats organisms face equips one to develop a strategy for minimizing the future impacts on the species and populations of concern. The first step towards the prevention of global extinction is the identification and documentation of species, ecosystems and landscapes, and the threats that face them.

Habitat loss

There is agreement that the greatest threat to both animal and plant species lies in the loss of habitat. In many instances organisms disappear from a system even before it has been totally transformed; this is usually the result of the disruption of the functional processes necessary to sustain, intact, the species complement. Although outright habitat destruction clearly has most impact, habitat degradation, as a result of selective harvesting (including logging), heavy grazing or seasonal burning, certainly has an effect by causing such things as loss of nest-sites, direct disturbance and alterations to the mix of plants species, including food plants. Additionally, there are many subtle pressures linked to habitat degradation that are poorly understood, as the effects may vary from species to species and even seasonally within species.

The factors causing habitat loss may vary within and between habitats. For example, in mesic areas, conversion to large-scale agriculture (eg, by fire, stocking rate, chainsaw, plough or bulldozer), clearance by small-scale farmers, large-scale planting or logging, infrastructural development (eg, buildings, dams, power lines and roads) and mining are the main agents of habitat transformation. In drier habitats, conversion to rangeland, exotic plantation, agriculture, overgrazing and excessive fuelwood collection are the principal direct causes of habitat loss and degradation.

However, the above mentioned agents are not in reality the underlying causes of habitat loss. Macro-economic problems faced by many developing countries are the driving force behind the conversion of habitats into apparently economically attractive land-use options as nations struggle to meet repayments on international debts. Many countries have been, and are, plundering their habitats or clearing them to reach mineral deposits in order to satisfy the insatiable economies of the developed consumer countries. In many cases, rural people have been displaced by dollar-earning commodity crops and are forced to wreak their own slash-and-burn devastation in marginal areas, thus exacerbating the over-arching problem of habitat loss. In addition to the dire economic circumstances faced by many developing countries, action to counter habitat loss is hindered by:

- domestic legislation (in particular with respect to land tenure);
- the absence of incentives for 'good practice';
- corruption among officials;
- a lack of political interest in conservation; and
- inadequate human and financial resources.

The consequence of such extensive processes of habitat loss is that the geographical distribution of many threatened species has shrunk so drastically that they survive in only a few fragments of their former ranges.

Over-exploitation

Populations of commercially viable species may decline rapidly if exploitation occurs at levels that exceed their reproductive potential. Even superabundant species can become extinct in a few years if exploitation is excessive. However, commercial exploitation is seldom, if ever, the sole cause of extinction.

The passenger pigeon (*Ectopistes migratorius*) showed a spectacular decline in numbers from billions in 1810, to around 200 million in 1870, to one captive female only 40 years later, and finally, extinction in 1919. This is frequently used as a classic example of overkill leading to extinction (King, 1987, and references therein). However, Bucher (1992) has recently presented a convincing counter-argument. He contends that the passenger pigeon became extinct primarily as a result of forest destruction and fragmentation, particularly in its northern breeding grounds. The combination of the loss of critical breeding habitat and the absence of social cooperation among the birds in food finding at low densities would have been enough to lead to the extinction of this species even without the killing of a single bird and despite the existence of considerable remaining forest.

Large harvests are not necessarily a sign of over-exploitation. More than 10 million waterfowl are killed by North American hunters in a regulated annual harvest, but, barring a few exceptions, this does not exceed the reproductive success of the species. Commercial exploitation features in neither academic nor textbook treatments of extinction (Hobbs and Mooney, 1998; Lawton and May, 1995; Novacek and Wheeler, 1992).

Introduced species

Biotic invasions are ongoing worldwide, the full extent of which has yet to be fully documented. Nevertheless, invasions have become recognized as a major factor forcing global environmental change (Vitousek *et al*, 1996). Perhaps one of the best examples of multiple threats to a single species posed by a variety of introduced organisms is the case of the Mauritius parakeet (*Psittacula echo*). This species suffers nest predation from crab-eating macaques (*Macaca fascicularis*) and rats, and competes for food and nest-sites with the introduced ring-necked parakeets (*P krameri*). Moreover, the last fragments of its native forest habitat are at risk from encroachment by introduced plants.

Pollution and pesticides

Pesticides are often cited as a major cause of the decline of natural insect populations and their use in restricted natural habitats must be viewed with concern. However, the effects of pesticide use, in particular the side-effects of widespread agricultural and forestry pesticides (which have sometimes been applied without concern for problems of drift and contamination of nearby non-target habitats), have been difficult to document.

Decreasing range size

Much of the current scientific and public concern over the extinction crisis revolves around the loss of species globally. Most of the benefits which biodiversity confers on humanity, however, are dependent on large numbers of populations of species, because each population ordinarily provides an incremental amount of an ecosystem's value or service (Hughes *et al*, 1997). Hughes *et al* presented a case in which population diversity and area correspond roughly in a one-to-one fashion over ecological time. Thus, when 90 per cent of an area is destroyed, about 90 per cent of the populations inhabiting the original area are exterminated (as opposed to roughly 50 per cent of the species as predicted by the species-area relationship).

A BIRD'S VIEW OF THREAT

Birds are the best known taxon. Collar *et al* (1994) produced the first comprehensive analysis of the threats facing them, using the revised IUCN criteria. It is thus useful, when making a quantitative estimation of the various threats that may erode global species diversity, to consider both birds in general and a selection of sub-divisions within the group.

A total of 1,111 bird species, 11 per cent of the world's avifauna, are threatened in one way or another (Collar *et al*, 1994). Additionally, a further 10 per cent of species are considered as potentially at risk. Overall, more than one-fifth of all bird species give some cause for concern in terms of the likelihood that they could be at risk of extinction. Of the 1,111 species considered to be threatened, 52 per cent

Table 2.1 *Breakdown of the relative importance of threats to various subsets of birds*

| | Number of species | Threat (%) | | | | | | | |
		Habitat loss	Small range	Subsistence hunting	Introduced species	Trade	Natural causes	Unknown	Other
Restricted-range species (Bibby *et al*, 1992)*	2,609	63.0	–	6.0	8.0	–	–	17.0	6.0
Threatened birds (Collar *et al*, 1994)*	1,111	51.9	23.2	7.6	5.8	2.6	3.3	2.5	3.1
Threatened African birds (this study)**	151	85.4	37.7	17.2	1.3	1.3	4.6	3.9	–
Threatened parrots (Juniper and Parr 1998)**	90	81.1	–	–	–	43.3	–	–	–

* giving equal weight where multiple threats are operating (ie total = 100%)
** giving overall contribution of each threat type (ie total = >100%)

are considered to be threatened chiefly by habitat loss and alteration, 23 per cent by small range or population, almost 8 per cent by subsistence hunting, 6 per cent by introduced species, while commercial trade threatens less than 3 per cent of such species (see Table 2.1). A similar picture emerges from a brief survey of some particular categories of birds.

Restricted-range species

Some 20 per cent of all birds are confined to just 2 per cent of the earth's land surface. Bibby *et al* (1992) collected distribution records of all bird species with breeding ranges below 50,000 sq km (ie, the size of Denmark or Costa Rica). They found that of 2,609 restricted-range species, 63 per cent are threatened by habitat destruction, 8 per cent by introduced species, 6 per cent by hunting (mainly for subsistence or because they are pests), 17 per cent by unknown factors and 6 per cent by other factors (including trade, disturbance, pesticides, poisons, fisheries and natural causes, etc) (see Table 2.1).

Threatened African species

Africa provides a suitable example for a continental perspective of the threats facing birds as it not only holds a significant proportion of the total number of species, but its biota are also subjected to every conceivable form of threat. Furthermore, the use of wildlife features strongly in the cultures of most African nations and in many cases the use of wildlife presents one of the few practical means of earning a livelihood. Thus, while the commercial use of wildlife may be viewed with circumspection in the developed world, it is a rare novelty to be able to take this position in Africa.

Of 1,700 bird species occurring in Africa, 151 species (or almost 9 per cent) are considered to be threatened (ie, excluding those species considered near-threatened) (Collar *et al*, 1994). Of these, 85 per cent of species are threatened by habitat loss or degradation, 18 per cent by subsistence hunting, and 11 per cent by pollution and pesticides. Commercial trade threatens only 2 species (ie, less than 2 per cent), only one of which is restricted to the continent (see Table 2.1). So, while a higher proportion of African species are threatened by subsistence hunting, habitat loss is again the most important threat, with commercial trade being much less significant.

Parrots

Among birds, parrots are the group most subject to commercial trade. A wide range of human societies, from the indigenous people of the rainforests to the technologically advanced societies of North America and Europe, place a particular value on parrots as pets. This is, indeed, not a recent occurrence as parrots have featured as companions to humans almost as long as recorded history itself. Data collected under the obligations of signatories to CITES, and analysed by the IUCN's Wildlife Trade Monitoring Unit, reveal the scale and breadth of trade in parrots (see Mulliken, 1995). Between 1980 and 1992, 247 species of parrot were reported in international trade, with 156 of them traded over that period in volumes exceeding 1,000 birds annually. If one were to add to this the illegal trade and the trade within national boundaries, these figures would be likely to increase substantially. In the light of this one might expect parrots to be threatened by commercial trade and, indeed, this is the case.

The parrots at risk of extinction and those deemed near-threatened face a wide range of pressures. In general, however, the principal sources of threat arise from habitat loss and the collection of birds for live trade (Juniper and Parr, 1998). Of 90 species at risk of global extinction, 81 per cent are threatened by habitat loss or degradation, 43 per cent by trade in live birds, with at least 31 per cent threatened by both pressures. Less than 7 per cent are considered to be threatened by factors other than these two (see Table 2.1). It is useful to compare the threats to parrots in Australia and South America. Between them, they provide the home to almost two-thirds of all parrot species. Unlike South America, however, Australia has implemented a total ban on trade in parrots for almost forty years.

The New World harbours around 140 parrot species, of which no fewer than 41 species (29 per cent) may be considered at some risk of extinction (Collar et al, 1994). Of these, almost 90 per cent are threatened by loss of habitat and 41 per cent by commercial trade. Habitat loss is the sole threat to the persistence of 14 species (34 per cent); only one species (2 per cent) is considered threatened by trade alone and 16 species (39 per cent) from a combination of habitat loss and trade. Of the 55 parrot species that occur in Australia, ten species (18 per cent) are at some risk of extinction (Collar et al, 1994). Of these, all are at risk from habitat loss and 40 per cent from a combination of this and trade. Although a smaller proportion of the Australian parrot fauna is considered threatened when compared to that of the New World, the threat profiles of the parrots inhabiting these

Table 2.2 *Breakdown of threats facing parrot species of the New World (equivalent to the Neotropics) and Australia*

Threat	Equal weighting (%)		Absolute frequency (%)	
	New World *41 species*	Australia *10 species*	New World *41 species*	Australia *10 species*
Habitat loss and alteration	57.5	43.4	87.5	100
Trade (of live birds and egg collection)	16.6	17.5	41.5	40
Hunting (subsistence and persecution)	4.5	5.0	14.6	10
Introduced species	0.6	13.3	2.4	10
Other factors (disturbance, small range, natural causes)	20.8	20.8	48.8	60

Note: The species are classified as 'critical', 'endangered' or 'vulnerable' by Collar *et al* (1994). In this table threats have been given equal weight where multiple types have been identified, and absolute weighting according to the proportion of species affected by each threat.

regions are virtually identical (see Table 2.2). The higher percentage of threatened New World parrots can be ascribed largely to severe pressures on their habitat.

This survey of the threats both to birds in general and to particular categories of birds confirms the general picture of the threats to wild species that was outlined in the previous section. Habitat loss is by far the most important threat to birds and commercial trade is typically a serious threat to a much smaller number of species. A partial exception to this is provided by parrots. But even for parrots, where commercial trade is a significant threat, habitat loss is more important.

THE RESPONSE OF CITES

In view of the above analysis of the nature of the threats to species, CITES appears to be concentrated on just a small part of the problem. For CITES, as the name suggests, focuses on just one threat to species, that posed by international trade. It responds to the threat by attempting to halt or restrict that trade. Species can be placed on either Appendix I or Appendix II of the convention. International commercial trade in Appendix I species is almost completely banned and trade in Appendix II species is subject to regulation. However, as we have seen, for most threatened species commercial trade is not a significant threat. Habitat loss is far more important and CITES has no remedy for this problem. Even for the relatively small number of species for which commercial trade is a threat, habitat loss would often appear to be a greater threat. The fact that CITES is only addressing a small fraction of the threats to species might not matter so much were it not the case that CITES is perhaps the most important and well-known conservation convention. It helps to determine where conservation efforts are directed and if this convention does not address the most important threats, then something is awry.

There are other weaknesses within CITES. Even for those species where trade may be a threat, it is not clear that the convention always deals effectively with it. This is partly because imposing restrictions on international trade in wild species sometimes just has the effect of driving the trade underground. For example, substantial numbers of parrots change hands across international boundaries despite the existence of CITES. Some workers estimate this type of activity to be at least equal in magnitude to the legal trade. The comparison between Australia and the New World is instructive here. Even though Australia has implemented a total ban on the export of its parrots for almost 40 years, trade in live birds apparently poses as much of a threat to Australian parrots as it does in the New World. Additionally, species listed on Appendix II are sometimes captured in a country where a wildlife export ban exists and then moved to a country where documentation of local origin can be obtained, permitting 'legal' exportation. Furthermore, even when CITES is able to control international trade, domestic trade is left untouched. Harvests for international trade represent only a part of the actual offtake of wild species. In the case of birds, many are trapped and traded at the national level. These numbers will in some cases be higher than the numbers exported.

There is also some evidence that where species are traded internationally, the convention appears to be responding to something other than the degree of threat alone. Of 90 parrot species that are listed by Collar *et al* (1994) as being under global threat, 33 of these species appear in Appendix I of CITES. A further nine non-threatened parrot species are listed in Appendix I. All other parrot species, with the exceptions of the ring-necked parakeet, budgerigar (*Melopsittacus undulatus*) and cockatiel (*Nymphicus hollandicus*), are included in Appendix II. It thus appears that the listing of parrots on Appendices I and II respectively, might relate largely to the level of commercial trade and not to the degree of threat to the species.

Because of the power and significance of CITES, it has the effect of encouraging research and funding into the management of species that are of most commercial interest instead of species that are most endangered. It is a perpetual problem to find funding for species on Appendix I. But species on Appendix II, especially those threatened by a trade ban, absorb huge resources, as commercial interests and nation states endeavour to prove that the species can withstand trade and should not be moved to Appendix I. A case in point is that relatively large sums of money have been spent on the orange-winged Amazon (*Amazona amozonica*), which is a common parrot with a wide distribution, while almost no attention has been paid to the vinaceous Amazon (*A vinacea*), which is under extreme threat. Closer links between CITES and the CBD, which could source funds for critically endangered species on Appendix I, may be one way of tackling this problem.

Links with the CBD would also provide a way of placing CITES in a wider context where all the threats to species are dealt with, for the CBD explicitly recognizes that there are a wide range of threats to species, including habitat loss. At the moment CITES appears to be regulating the international trade in species with little attention being paid to how these activities relate to the conservation of biological diversity in general and species in particular. The activities and objectives of CITES need to be reassessed so that they correspond more closely to the actual threats faced by species.

REFERENCES

Bibby, C, Collar, N, Crosby, M, Heath, M, Imboden, C, Johnson, T, Long, A, Stattersfield, A and Thirgood, S (1992) *Putting Biodiversity On the Map: Priority Areas for Global Conservation*, International Commission for Bird Preservation, Cambridge

Bucher, E (1992) 'The causes and consequences of extinction of the Passenger Pigeon', *Current Ornithology*, Vol 9, pp 1–36

Collar, N, Crosby, M and Stattersfield, A (1994) *Birds to Watch 2*, BirdLife International, Cambridge

Hobbs, R and Mooney, H (1998) 'Broadening the extinction debate: population deletions and additions in California and Western Australia', *Conservation Biology*, Vol 12, pp 271–283

Hughes, J, Daily, G, and Ehrlich, P (1997) 'Population diversity: its extent and extinction', *Science*, Vol 278, pp 689–692

Juniper, T and Parr, M (1998) *Parrots: A Guide to the Parrots of the World*, Pica Press, Sussex

King, F (1987) 'Thirteen milestones on the way to extinction', in Fitter, R and Fitter, M (eds) *The Road to Extinction*, IUCN, Gland

Lawton, J and May, R (1995) *Extinction Rates*, Oxford University Press, Oxford

Mace, G and Collar, N (1994) 'Extinction risk assessment for birds through quantitative criteria', *Ibis*, Vol 137, S240–S246

Mulliken, T (1995) *Response to Questions Posed By the RSPB Regarding the International Trade in Wild Birds*, TRAFFIC International, Cambridge

Novacek, MJ and Wheeler, QD (1992) 'Introduction – extinct taxa', in Novacek, MJ and Wheeler, QD (eds) *Extinction and Phylogeny*, Columbia University Press, New York

Vitousek, P, D'Antonio, C, Loope, L and Westbrooks, R (1996) 'Biological invasions as global environmental change', *American Scientist*, Vol 84, pp 468–478

Part II

CITES IN PRACTICE

Chapter 3

When CITES Works and When it Does Not

R B Martin

INTRODUCTION

Has CITES been successful? If the evidence is inconclusive then can we specify the circumstances in which it is likely to be successful? And what are the lessons to be drawn from this? This chapter examines these vexed questions. It begins by considering the direct evidence that CITES has helped to conserve wild species. It then discusses the sort of factors that are likely to make CITES successful. Finally, some reforms are proposed.

HAS CITES WORKED?

CITES came into force in 1975 and is, in principle, a simple convention. Species can be listed in one of three appendices. Species listed in Appendix I are supposed to be on the brink of extinction and may not be traded commercially; species listed in Appendix II are not yet threatened with extinction but may become so if trade is not controlled;[1] and species are listed on Appendix III when a party is regulating a species within its jurisdiction and it does not want the species to be traded internationally without its express permission.

In addressing the question of whether CITES has worked, numerous claims have been put forward for the value of the convention. It has been said that it has led to a greater awareness of conservation issues;

---

[1] Appendix II also includes species which resemble those listed on Appendix I ('look-alikes') on the assumption that trade in such species, even if they are not threatened, also needs to be controlled.

that it has forced the Parties to the convention to strengthen their domestic implementation of conservation measures; and that it has reduced demand for the products of endangered species. There may be an element of truth in all these assertions. Nevertheless, the effects in question are, at most, byproducts of the convention and they could have been achieved through other means.

There is only one direct test of the performance of CITES – has the convention improved the status of the species of wild fauna and flora that it sets out to protect? However, there is no clear cut answer to this question. There are no species whose numbers have increased so dramatically after being listed on the CITES appendices that the improvement is obvious. So, in attempting to answer the question accurately, considerable care is needed. In order to show that an improvement was caused by a listing on Appendix I of CITES one would need data on the global species population that demonstrated two things. Firstly, it would have to be established that the population was declining to the point that it was threatened with extinction immediately before it was listed on Appendix I and that the decline was definitely caused by unsustainable international commercial trade. Secondly, it would have to be shown that the population increased in numbers after the listing to the point where the population could be deemed to have recovered and could be transferred to Appendix II so that it could once more be traded commercially on a sustainable basis. Moreover, the figures would have to demonstrate that the increase was due to the listing and not to other factors such as an intrinsic population increase or improved law enforcement.

It has not yet been shown that there are any species that have satisfied these criteria. Trexler has claimed that there is no measurable evidence that CITES has benefited any species at all (Trexler, 1990). In a recent review of the effectiveness of CITES commissioned by the Parties to the convention, the consultants examined the status of 12 selected species and were able to conclude that only two *appeared* to have improved as a result of their listing on CITES appendices (Environmental Resources Management, 1996). This is telling stuff. If the convention is benefiting species then, even after careful study, it has not been demonstrated.

One of the two species that the consultants concluded had improved in status as a result of the policies of CITES was the Nile crocodile (*Crocodylus niloticus*). However, in a well-prepared submission to the consultants carrying out the review, the Crocodile Specialist Group of the IUCN Species Survival Commission analysed the various factors which contributed to the recovery of the Nile crocodile populations. They contended that the status of the Nile crocodile improved not as a result of applying the standard CITES medicine, but as a consequence of departing from that prescription. Specifically, it was only when CITES shifted from a policy of restricting trade to one of promoting the sustainable use of crocodiles that crocodile numbers increased.[2] The direct evidence for whether CITES has been successful is, at best, inconclusive. Nevertheless, this does not mean that nothing more can be said. It is still possible to pose a more general question about the *sorts* of circumstances in which CITES is *likely* to be successful. In addressing this question one is enquiring into the causal factors – including the policies of CITES themselves – that determine the fate of species. The experiences of the last 25 years provide some help in answering this question.

WHEN IS CITES LIKELY TO WORK?

In considering the factors which are likely to influence the success of CITES, it is useful to make a rough distinction between institutional and policy factors. I will begin with the former.

CITES is an international agreement which depends on the Parties to implement its decisions. This has implications for when CITES is likely to work well. It suits Parties where wildlife control is strongly centralized and efficiently managed, where citizens have legal rights to use wildlife only as permitted by government agencies and where this central control is popularly accepted.[3] In such systems the national bureaucracy will be well placed to implement CITES controls effectively. Moreover, CITES will be most effective when it works in concert with national states and not against them; that is, when it aids the Parties' own law enforcement efforts to control illegal or excessive trade. It is best treated as an extra facility under which any Party can invoke the assistance of law enforcement agencies of other Parties in improving the implementation of its own policies. This presupposes a high degree of mutual respect for the sovereign rights of nations and tolerance of a wide variation in approaches to conservation issues.

---

[2] The case of crocodiles is discussed further by H Kievit in Chapter 8 of this book.

[3] This statement should not be taken to endorse strong, centralized institutions as the appropriate national strategy for wildlife conservation and management. Indeed, investment in government agencies to carry out conservation is unlikely to be as effective or cost-efficient as devolving power to local institutions.

Where these conditions are not satisfied, however, CITES is un-likely to work. This will be the case where control of wildlife is not centralized or not popularly accepted or where the state bureaucracy is weak and inefficient. In these circumstances no amount of controls at the international level can rectify the weaknesses of state agencies. Moreover, if the fate of wildlife is effectively determined by rural people rather than by government agencies, and if states are repre-sented by government officials who may have little interest in or knowledge of rural people, then CITES is not likely to provide an effective forum for dealing with wildlife issues. Additionally, the ef-fectiveness of CITES will be undermined if the aims of all the par-ticipating Parties do not coincide. There is a common misconception that CITES provides protection to species. In fact, protection can only be achieved by law enforcement agencies and citizens of range states. CITES provides no short cuts and its effectiveness will be se-verely diminished if it is working against some national states. Thus, CITES will be unlikely to work when it is used as a mechanism to en-able certain Parties to impose their perceived conservation solutions on other Parties. Such an attitude is contrary to the spirit needed to form conventions. The voting system within CITES is unique in that it allows Parties who bear no financial costs for the protection of spe-cies which occur in other Parties' countries to, nevertheless, take de-cisions with financial implications for those range states.

In light of this, Article XIV of the original treaty, which allows Parties to take stricter domestic measures, is very important. It would appear to be reasonable for range states to make occasional use of the provision of Article XIV for species which might, for example, be heavily traded by their neighbours but for which they wished to ap-ply more rigorous controls than provided for in the treaty. But, with the passage of time, it is consumer states that have made most use of this provision. By imposing controls on the import of wild species that are stricter than those agreed by CITES they have been able to impose their own conservationist agenda on the range states, often without prior consultation. This tends to nullify the purpose for which states come together to form conventions.

As regards the functioning of CITES itself, the meetings of the Conference of the Parties to CITES are probably the best organized meetings of any convention. Documents are submitted a minimum of 150 days before the meeting, the agenda is set well in advance, every session starts and finishes exactly at the stated times, rules of procedure are strictly adhered to and documentation arising from each day of the meeting is distributed the following morning. The

Secretariat deserves full credit for this organization. But this, in itself, does not ensure success. The quality of the delegates is also important. CITES meetings worked extremely well when the delegates were senior technical officials functioning in their appointed capacities as Scientific and Management Authorities. Up until 1987, plenary and committee sessions were characterized by fast, intelligent debate among independent delegates who took no prior positions on issues and were capable of assessing the arguments presented.

Unfortunately, these standards no longer obtain. The positions taken by many nations are now motivated by political considerations unrelated to conservation and they are decided in advance of the meeting so that debate is futile. Many of the delegates are junior bureaucrats without authority or experience. They come to the meeting without having done a thorough job of preparation and hence are incapable of making significant interventions. Essential background documents, such as the reviews from IUCN and TRAFFIC (the wildlife trade monitoring programme of WWF and IUCN) on listing proposals, are not read or are ignored. Few Parties take any interest in proposals involving species which occur outside their boundaries.

Turning to issues of policy, CITES works best when the positive role which trade can play in wildlife conservation is acknowledged and when it is prepared to use its appendices flexibly. The beneficial effect of trade has been recognized for some time in a number of countries. For example, in many southern African states there is a significant move to promote wildlife management, in all its diverse forms, as a primary form of land-use – both as a means to improve human livelihoods and as a way of securing wildlife habitat.

CITES has shown some signs of recognizing this point. In 1992 a resolution was adopted that recognized that trade could promote wildlife conservation. More substantively, endangered species of crocodilians benefited from commercial trade when CITES transferred them from Appendix I to Appendix II. This was achieved through quota systems which provided incentives for Parties to improve their management. This is an example of CITES at its most effective. The Parties showed flexibility and actively sought solutions to allow trade that would benefit the species, regardless of the degree to which it was endangered. What made this solution possible in this case was the high commercial value of crocodile products, which allowed substantial investments in the management of the species. So CITES reduces illegal trade indirectly (but effectively) when the emphasis is placed on promoting a controlled, legal trade.

Where the high commercial value of products has been ignored and attempts have been made to destroy potential legal markets, it has seldom resulted in any lasting gains for the species involved (Dublin *et al*, 1995). Trade bans generally have the effect of driving trade underground and removing all means of monitoring. The black rhino (*Diceros bicornis*) is a classic example of a case where CITES has failed and will always fail if it does not take note of the realities of a given situation. In this instance the Parties have stuck rigidly to the provisions for no commercial trade in Appendix I species and the range states concerned have had few funds to improve the status of the species and little incentive to do so. Thus, the trade ban failed. The magnitude and nature of the demand was underestimated; the costs of the protection system pursued by CITES were too high; the illegal trade could not be controlled through the limited mechanisms available; and there were no incentives, such as existed for the Nile crocodile, to conserve the species.

What limits CITES is the binary system of two appendices and the provisions of Article III, which debar trade for Appendix I species when that trade is primarily for commercial purposes. The dilemma in which Parties find themselves is whether to give precedence to the endangered status of a species and list it on Appendix I, or whether to ignore its endangered status and list it on Appendix II so that beneficial trade can take place.

The importance of incentives for conservation also helps to explain why captive breeding is often not effective as a means of conservation. At one stage in the history of CITES captive breeding was seen as a valuable conservation tool which could remove the pressure on wild species by providing an alternative harvest. However, in addition to the pollution, disease and genetic problems which can arise in captive rearing situations, the great disadvantage is that, having obtained a founder population from the wild, little reason remains to reinvest in the conservation of wild populations (Luxmoore and Swanson, 1992). CITES should actively seek to encourage sustainable wild harvests. The banning of all trade, other than from captive breeding programmes, does not act as an incentive to conserve wild populations.

As it is not always easy to predict the effects of a particular policy on wild species, adaptive management is important and CITES would work well if this approach to its implementation was adopted. The use of species would not be subject to a prerequisite of expensive surveys and research, the costs of which invariably fall on the range state and the results of which cannot establish how a population will respond to exploitation. Instead, the monitoring of the harvested population would provide the data needed to improve management and ensure sustainability. At present a listing on Appendix I need not be accompanied by any monitoring or improvements in management and there is a tendency for certain developed countries to demand certainty based on 'scientific principles' before trade can be contemplated. This demonstrates a limited understanding of the tools science has to offer. In such situations the attempt to make predictions based on a limited understanding of the parts of the whole and on the assumption of a stable environment is an inferior form of science (Holling, 1993). Dublin has remarked that this approach is likely to result in the greatest divisions between developed and developing countries in the coming years (Dublin, 1996).

In all of this, it is important that species are listed for the right reasons. CITES is concerned with species threatened by international trade and its appendices should reflect this. However, there is over-representation of so-called 'charismatic' species attractive to human beings and under-representation of some of the vertebrate orders with many endangered species, such as fish. If the Parties move rapidly to apply the new criteria for listing species on the appendices, this could rectify the present imbalance.[4] Appendices which are over-burdened with species that are not threatened weaken the credibility of CITES and can lead to a dilution of conservation efforts and funds. It is also a feature of CITES that when a species population in one range state is threatened it is often listed on Appendix I, regardless of its secure global status. In such cases it should simply have been listed on Appendix III by the affected range state.

WHAT CHANGES ARE NEEDED TO MAKE CITES WORK BETTER?

In a perfect world, if every nation state possessed strong, competent agencies (wildlife, forestry, fisheries, customs, police etc) to conserve and manage its fauna and flora sustainably and to minimize illegal trade, there would be no need for any convention to regulate international trade. But it is not a perfect world. There is a wide variation in the capacity and political will of different states to conserve natural resources. As a result, there is widespread mistrust and dissatisfaction

---

[4] New criteria were proposed at the 8th CITES Meeting in Kyoto (1992) and introduced at the ninth CITES Meeting in Fort Lauderdale (1994).

about how the responsibility to conserve what is perceived to be a global heritage is distributed between individual nation states. So, a convention is needed. Given our knowledge as we approach the turn of the century, what changes should be made to the existing CITES convention?

In the light of the above discussion, three recommendations can be made. Firstly, the existing system of two appendices should be replaced with a single appendix on which are listed all species of international concern and the focus should be on appropriate management programmes for each species. This may entail introducing a quota system for each species such as has been used under CITES for crocodiles. Quotas could vary from zero upwards, thus retaining the possibility of imposing a trade ban on any species. The value of the quota system lies in the fact that Parties are obliged to take into account their national utilization and this results in an overall improvement in management, a feature which has so far eluded CITES because of its limited focus on international trade. Moreover, an amended treaty would provide a good basis for involvement in the sustainable use of timber and fish species. These are not the business of CITES at present because the articles of the convention limit it to consideration of species threatened with extinction.

Secondly, and relatedly, the treaty should take more account of the different circumstances in different countries. Such an approach is consistent with the decentralized strategy, which IUCN's Sustainable Use Initiative is adopting to cater for the variation in approaches to sustainable use in each region of the world. Given the current situation with CITES, where the United States, the European Union and Australia are using stricter domestic measures to differentiate among imports from individual states, this would not be a new phenomenon for CITES. It implies a departure from concerns about the global status of species but, since that has already happened in a *de facto* sense, it is not a departure from the present status quo.

Finally, if reformed in this way, it would make sense to place CITES under the larger umbrella of the CBD. The time may have come to reconstitute CITES as a new protocol under the CBD, incorporating the recommendations made here.[5]

---

[5]   The relationship between CITES and the Convention on Biological Diversity is discussed further in Chapter 11.

REFERENCES

Dublin, H, Milliken, T and Barnes, R (1995) *Four Years After the CITES Ban: Illegal Killing of Elephants, Ivory Trade and Stockpiles*, IUCN Species Survival Commission Report, Gland

Dublin, H (1996) 'North–South dissonance on consumptive use of charismatic megafauna', in *The Proceedings of a Pan African Symposium on Sustainable Use of Natural Resources and Community Participation*, IUCN ROSA, Harare

Environmental Resources Management (1996) *Study on How to Improve the Effectiveness of CITES*, Final Report to the Standing Committee of CITES, Lausanne

Holling, C (1993) 'Investing in Research for Sustainability', *Ecological Applications*, Vol 3, No 4, pp 552–555

Luxmoore, R and Swanson, T (1992) 'Wildlife and Wildland Utilization and Conservation', in Swanson, T and Barbier, E (eds) *Economics for the Wilds*, Earthscan, London

Trexler, M (1990) 'The Convention on International Trade in Endangered Species of Wild Fauna and Flora: Political or Conservation Success?', PhD Dissertation, University of California at Berkeley

Chapter 4

Precaution at the Heart of CITES?

Barnabas Dickson

INTRODUCTION

In 1994 CITES explicitly adopted the precautionary principle for the first time. It might have appeared that CITES was simply catching up with the rest of the environmental world. From the late 1980s onwards the precautionary principle was included in an increasing number of international declarations and agreements. The conclusive stamp of approval came in 1992 when it was endorsed in Principle 15 of the Rio Declaration on Environment and Development. However, it can be argued that the principle was implicitly present in the original CITES treaty, signed in 1973.

The first aim of this chapter is to identify the ways in which CITES can be said to have endorsed the precautionary principle both before and after 1994. The picture that emerges is a confused one. The second, more ambitious aim is to assess what role the precautionary principle might legitimately have within CITES. This can then be used to evaluate the versions of the precautionary principle that are found in CITES.

These two tasks are made more difficult by the lack of clarity that surrounds the precautionary principle. As well as being incorporated in international law, the principle figures prominently in popular discussions of environmental issues. But despite the widespread currency of the principle, there is no agreement on its precise import or even on its formulation. I will begin by tackling these issues.

THE FORMULATION AND ROLE OF
THE PRECAUTIONARY PRINCIPLE

Many discussions of the precautionary principle are vitiated by a lack of clarity about what the principle itself states and the role it might play in determining policy. In an effort to shed some light on these issues, this section makes two distinctions. The first is between two different versions of the precautionary principle and the second is between two sorts of role that a practical standard, such as the precautionary principle, might play in determining policy decisions.

It is a remarkable fact about the formulations of the precautionary principle that appear in a wide range of international environmental agreements that almost no two formulations are identical. Nevertheless, it is possible to make a broad distinction between two versions of the principle. The first version was incorporated in agreements that mainly addressed marine pollution. The formulation that appears in the 1989 report of the Nordic Council's International Conference on the Pollution of the Seas will serve as an example of this version. The report speaks of:

> *The need for an effective precautionary approach, with that important principle intended to safeguard the marine eco-system by, among other things, eliminating and preventing pollution emissions where there is reason to believe that damage or harmful effects are likely to be caused, even where there is inadequate or inconclusive scientific evidence to prove a causal link between emissions and effects.*

According to this version the precautionary principle recommends taking action against a practice which may be causing damage to the environment even if there is no proof of a causal link. It can be described as the action-guiding version of the principle. It says that something should actually be done about the emissions even before it is proved that they are causing harm. The action-guiding version can be found, with some variations, in the Ministerial Declaration of the second International Conference on the Protection of the North Sea (1987) and in the recommendations of the Paris Commission established by the parties to the Convention for the Prevention of Marine Pollution from Land Based Sources (1989).

Subsequently, a rather different version of the principle was included in environmental agreements that had a wider application.

The formulation that appears in Principle 15 of the Rio Declaration on Environment and Development (1992) is typical:

> *In order to protect the environment, the precautionary approach shall be widely applied by States according to their capabilities. Where there are threats of serious or irreversible damage, lack of scientific certainty shall not be used as a reason for postponing cost-effective measures to prevent environmental degradation.*

This version of the precautionary principle, unlike the action-guiding version, does not recommend action in situations of uncertainty. It simply stipulates that the absence of scientific certainty should not be used as a reason for not taking action. It can be characterized as the deliberation-guiding version of the principle. It offers guidance on how one should deliberate in situations of uncertainty.

The second distinction I want to make is essentially the same as the one made by Ronald Dworkin in the course of his critique of legal positivism (Dworkin, 1977, pp 22–28). Dworkin makes a distinction between legal principles and legal rules. Both are practical standards that point to a certain sort of action as something that is to be done in certain specified circumstances. But they differ in the nature of the direction they provide. A rule requires that the action be performed if the specified circumstances obtain. If the action is not performed then the rule has been broken. In contrast, a principle does not require that the action be performed. It provides a reason for performing the action, a reason that should be taken into consideration. But it is a reason that can be outweighed by other principles. If a principle is outweighed in this way then it has not been violated in the same way as a rule has been broken if the guidance it offers is not followed.

This distinction of Dworkin's can be usefully applied in the field of policy making. It distinguishes between two different ways of interpreting a practical standard. If the standard is interpreted as stipulating what is to be done in certain circumstances and does not admit of exceptions then it is being treated as a rule. If it is interpreted as stating a reason that should be considered in certain circumstances, but that need not always prevail, then it is being treated as a principle. In the context of the present discussion the key question is whether the precautionary principle is to be interpreted as a principle (in Dworkin's sense) or as a rule. As this last sentence indicates, there is a possibility of terminological confusion here, for the question is whether the precautionary principle should be interpreted as a principle or a rule. However, as long as it is borne in mind that there are two senses of principle (a looser sense in which the precautionary principle is indeed a principle and a more specific sense, derived from Dworkin, in which it is an open question whether the precautionary principle is a principle or a rule) the scope for misunderstanding should be minimized.

The rule/principle distinction is of most relevance when it is applied to action-guiding versions of the precautionary principle. For depending on whether it is intended as a rule or a principle, the action-guiding version will require action to be taken or will simply offer a reason for action to be taken. The difference in the policy implications is quite considerable. Since the deliberation-guiding version of the precautionary principle only offers guidance on how to deliberate, the difference between interpreting it as a rule and as a principle is less dramatic (although still important in some circumstances). We can now deploy these two distinctions in the survey of the precautionary principle in CITES.

THE PRECAUTIONARY PRINCIPLE IN CITES

It can be argued that the central articles of the original CITES treaty embody a version of the precautionary principle. These are the articles that specify when species should be listed on Appendix I or Appendix II of the treaty and what the consequences of such listings are. Article II(1) states 'Appendix I shall include all species threatened by extinction which are or *may be* affected by trade' (emphasis added). Article II(2)(a) states that Appendix II shall 'include all species which although not necessarily threatened with extinction *may become so* unless trade in specimens of such species is subject to strict regulation' (emphasis added). Articles III and IV then specify the implications of a listing on these appendices. An Appendix I listing requires an almost complete ban on international trade in the species. An Appendix II listing requires some regulation of the trade.

Since a listing on the appendices is required even when it is not certain that trade may be harming the species and since such a listing then requires that action be taken against the threat, one can treat these articles as embodying an action-guiding version of the precautionary principle. Moreover, it seems clear that this version is, in Dworkin's sense, a rule rather than a principle. That is, the articles are not to be interpreted as suggesting that the Parties *consider* taking action even when the evidence that trade is harming species is inconclusive. Rather, they stipulate that action, of the specified sort, is to be taken whenever a threatened species is affected by trade or whenever a species may become threatened if trade is not strictly regulated. The treaty does not allow for any exceptions. Although CITES was implicitly precautionary from its inception, the picture was complicated when it came to explicitly endorse the precautionary principle. This happened in 1994 at the Fort Lauderdale Conference of the Parties, when Resolution Conf 9.24 was passed. This resolution provides detailed criteria for making changes to the lists of species on Appendix I and II. It attempts to improve on the imprecise criteria that were supplied by the original articles and subsequent resolutions.

The Resolution explicitly endorses the precautionary principle in two places. The second of these endorsements is the fuller one and it says that the Conference of the Parties:

> RESOLVES that when considering any proposal to amend Appendix I or II the Parties shall apply the precautionary principle so that scientific uncertainty should not be used as a reason for failing to act in the best interest of the conservation of the species (Resolution Conf 9.24, 1994).

This formulation is clearly an instance of the deliberation-guiding version of the precautionary principle. It does not specify that a species should be listed despite uncertain evidence. It simply says that the uncertainty should not be used as a reason for not acting in the best interest of the species.

There is, therefore, a sharp difference between the implicitly precautionary approach found in the original articles and the version of the precautionary principle that was explicitly endorsed by CITES. However, the picture is actually more complex than that. Much of the substance of Resolution Conf 9.24 appears in the annexes to it. The resolution itself stipulates that these annexes are 'an integral part of this Resolution'. Annexes 1 and 2a attempt to fill out the criteria that appear in Articles II(1) and II(2)(a) of the original convention for listing species on Appendices I and II. They are faithful to the spirit of those articles. For, in the detailed criteria they offer, they also implicitly endorse an action-guiding, rule-like version of the precautionary principle. For example, Annex 2a stipulates that a species should be included on Appendix II if it is 'known, inferred *or projected* that the harvesting of specimens from the wild for international trade has, *or may have*, a detrimental impact on the species'.

The emphasized words illustrate how action is required even in the absence of certainty. Annex 1 and the rest of Annex 2a contain other instances of similar wording. While these formulations may make the annexes consistent with the original treaty, they are also in some tension with the explicit version of the precautionary principle that is endorsed in the same resolution. So, the overall picture is confused. CITES offers contradictory recipes for responding to uncertainty.

EVALUATING THE PRECAUTIONARY PRINCIPLE IN CITES

The confusion about which version of the precautionary principle CITES endorses makes an assessment of what role the precautionary principle might legitimately play in CITES especially important. This assessment requires some discussion of the issues that confront any agency, such as CITES, that is attempting to devise a conservation policy for wildlife.

Faced with a decline in the numbers of a species, the question of what is the best policy to adopt in order to halt a further decline is a complicated one. A number of different sorts of consideration are likely to be relevant. One consideration concerns the identity of the processes that are causing the decline. CITES assumes that international trade is an important threat. But, as has become apparent in the last 25 years, and as is argued in other chapters in this book, there are other processes that pose a more serious threat to many species. Habitat loss is the most prominent of these. Clearly the nature of the threat has an important bearing on what is the best policy for halting the decline. A second consideration concerns the efficacy of the policies that might be adopted to cope with the threatening process. Habitat loss might be correctly identified as a threat but the creation of legally protected areas would not be an effective response if, in practice, the protection cannot be enforced. A third consideration concerns the unwelcome aspects of the policy. Even if the creation of protected areas did work, in the sense of preventing species depletion, it might also involve dispossessing rural people of land they depend on. This latter consideration would count against that policy. Considerations of this type are frequently present in countries of the South, where there is often the potential for conflict between the needs of people and the needs of wildlife. Equally clearly, the weighing up of these considerations will involve directly addressing evaluative questions about the relative importance of conservationist and social concerns.

A fourth, rather different sort of consideration that will affect the assessment of conservation policies concerns the varying degrees of uncertainty that may surround the issue. The seriousness of particular threats may be unknown and that could provide a reason for not taking action against the threat until more was learnt.

All these four types of consideration are likely to be relevant in determining which conservation policy is the best overall. Significantly, the precautionary principle, as a principle that offers guidance on what to do in the face of uncertainty, addresses only the fourth consideration. But in so far as the principle requires a readiness to take action even when there is uncertainty about the threats, then it would seem to offer sensible guidance. For if one waited until the evidence was more clear cut, then it may be too late; the species might have been lost. On the other hand, since the precautionary principle addresses only one of the relevant considerations, it cannot, on its own, determine what is the best policy, all things considered. Only in conjunction with other considerations about the sources of the threat, the efficacy of different policies for dealing with it, and the presence or absence of countervailing reasons, will the precautionary principle determine the best policy.

This allows one to draw a conclusion about the proper role of the precautionary principle in a convention such as CITES. The legitimate function of the principle is to highlight the importance of a readiness to respond to an apparent threat to species, even when there is some uncertainty about the seriousness of the threat. This is a significant role, especially given that some degree of uncertainty about threats is endemic in the field of wildlife conservation.

This view of the legitimate role of the precautionary principle can be used to evaluate the versions of the principle endorsed by CITES. The first such version is the one implicitly present in both the articles of the convention and in the annexes to Resolution Conf 9.24. This is a rule-like, action-guiding version. It requires restrictions on trade whenever a threatened species may be affected by trade or if a species may become threatened if trade is not strictly regulated. In the light of what has just been said about the range of factors that determine what is the best policy overall, this version looks misconceived. It does not allow room for the consideration of other processes that may be threatening species besides trade; or of other policies that might be more effective in conserving species; or of the other reasons that might exist for not restricting trade. By following this implicit version of the precautionary principle the Parties to CITES could be led to adopt a policy of restricting trade even though trade was not a significant threat and there were other policies that would do more to protect species.

The deliberation-guiding version of the precautionary principle that is endorsed by CITES in Resolution Conf 9.24 is not open to the criticism that it commits the Parties to placing trade restrictions on a species even when this would not be the best policy overall. Since it is a deliberation-guiding version it does not require any specific policies to be adopted. However, two aspects of this formulation might be questioned. On the one hand, if it is interpreted as a rule, then the requirement that scientific uncertainty should never be used as reason for failing to act might still be regarded as too strong. It seems reasonable to hold that if the evidence that a threat existed were *very* slim, this would count as a reason for postponing action. On the other hand, in cases where there is a significant amount of evidence of a threat, the constraint that this version places on the Parties is a very weak one. It simply states that uncertainty should not be used as a reason for inaction. It might be thought that the legitimate role of the precautionary principle would be better captured if the deliberation-guiding version stated that even in the absence of certainty the Parties should give serious consideration to action to protect the species.

In effect, these criticisms of the versions of the precautionary principle adopted by CITES imply that the Parties would do better to endorse a principle-like, action-guiding version or a reformulated deliberation-guiding version. Neither of these would rule out consideration of factors that are relevant to determining what is the best policy overall and both would draw attention to the importance of taking action to protect species even in the absence of certainty about the threats. It might be objected that both of these recommended versions of the precautionary principle would be virtually empty of content. For neither version would require a specific response to threats from the Parties. However, the main thrust of the argument here has been that it would be an abuse of the precautionary principle to treat it as mandating specific policy responses. No legitimate version of the precautionary principle can do this on its own. It is only in conjunction with other assumptions including, perhaps, contentious evaluative claims, that it will determine a specific policy. One of the temptations of the principle is that it seems to offer an apparently neutral principle that can be used to sidestep important evaluative questions. It does not. Indeed, the idea that the principle should lay down in advance the best course of action is one that runs contrary to the spirit of the precautionary principle. For given that we are often uncertain about both the precise nature of the threat to species and the best response to those threats, it is important that policy makers remain flexible and open to any new knowledge that may become available. The legitimate interpretation of the precautionary principle is one that acknowledges that it does contain an important message about action in the face of uncertainty, but also recognizes that other, quite distinct considerations help to determine the best policy overall.[1]

REFERENCES

Dickson, B (1999) 'The Precautionary Principle in CITES: A Critical Assessment', *Natural Resources Journal*, Vol 39, pp 211–228

Dworkin, R (1977) *Taking Rights Seriously*, Duckworth, London

---

[1]   The issues discussed in this chapter are dealt with in more detail in Dickson (1999).

Chapter 5

The Significant Trade Process: Making Appendix II Work

Robert W G Jenkins

INTRODUCTION

More than 140 countries are now signatories to the CITES. There is little doubt that it has become the most important and influential conservation treaty now in force. It was the product of a growing concern on the part of the international community about the rate at which wild species were declining. The general feeling was that the main threat came from the unregulated international trade in wildlife and that the solution was to create a legal framework to regulate that trade. When the convention came into force in 1975 it established a regulatory system at the heart of which are two appendices on which species can be listed. Appendix I is for species that are considered to be threatened with extinction. These species can not be traded for commercial purposes. Appendix II is for those species which might become threatened if trade is not controlled. Trade in Appendix II species is subject to some regulation.

After nearly 25 years of operation, and as we approach the next millennium, it is worth reflecting on the extent to which CITES has achieved the objectives that drove the international community to adopt it. This chapter addresses this issue by focusing on Article IV of the treaty. This article sets out the circumstances in which Appendix II species may be traded. It is thus a crucial element in the operation of the convention. In the early years of CITES Article IV was not applied effectively and this exposed flaws in the overall structure of the convention. Nevertheless, more recently CITES has proved sufficiently flexible to adopt a new mechanism governing the implementation of Article IV. Patching the convention in this way has given CITES the opportunity to become an effective tool for conservation.

ARTICLE IV

Article IV specifies when Appendix II species may be traded. Article IV(2)(a) states:

> The export of a specimen of a species included in Appendix II shall require the prior grant and presentation of an export permit. An export permit shall only be granted when the following conditions have been met:
> (a) a Scientific Authority of the State of export has advised that such export will not be detrimental to the survival of that species.

Thus, Article IV(2)(a) explicitly requires the Scientific Authority of an exporting State to determine the extent to which a population of a particular Appendix II-listed taxon can sustain an export trade. Some guidance on achieving this requirement is provided to the Parties by Article IV(3) which specifies the types of actions that must be undertaken by the Scientific Authority of an exporting Party. This clause states:

> A Scientific Authority in each Party shall monitor both the export permits granted by that State for specimens of species included in Appendix II and the actual exports of such specimens. Whenever a Scientific Authority determines that the export of specimens of any species should be limited in order to maintain that species throughout its range at a level consistent with its role in the ecosystems in which it occurs and well above that level at which that species might become eligible for inclusion in Appendix I, the Scientific Authority shall advise the appropriate Management Authority of suitable measures to be taken to limit the grant of export permits for specimens of that species.

In principle, the provisions of Article IV(3) constitute a mechanism to ensure that the removal of Appendix II plants and animals for export is undertaken in quantities that are sustainable and not detrimental to the long-term conservation of the wild population. However, in practice these provisions have not been applied in an effective way. A large part of the problem has been that the implementation of Article IV relies entirely on the will and capacity of the exporting countries. It is they alone who determine in what quantities species may be exported. There is no role either for the importing countries or for the international community. Thus, the extent to which the provisions have been adhered to has varied considerably with the legislative, administrative and technical capacities of the exporting country concerned. The result has been that Appendix II species are often exported at unsustainable levels. The failure of CITES to ensure that the trade in these species is sustainable has been very significant, for reasons discussed in the next section.

THE IMPORTANCE OF THE FAILURE OF ARTICLE IV

The most immediate result of the failure to ensure that Appendix II species are only traded at sustainable levels is that many such species have become more endangered. Consequently, in the absence of any mechanism to improve the implementation of Article IV the Parties have, understandably, turned to the only other option offered by the convention, namely the transfer of the endangered species to Appendix I. Indeed, the early years of CITES (1975–1989) saw large numbers of wild animals being transferred from Appendix II to Appendix I. This is not to say that all transfers to Appendix I have occurred because of the failure to implement Article IV. An ethical or ideological opposition to any commercial use of a species has sometimes been a factor. Such opposition has typically come from the non-range states of the developed world and this has tended to polarize opinion between developed and developing countries. The latter have claimed that species are listed in Appendix I in order to provide wealthy industrialized countries with green credentials at no cost to themselves. But whatever the reason for the transfer of species to Appendix I, its significance is increased by the difficulty of moving species in the reverse direction. The Conference of the Parties has traditionally adopted caution when considering proposals to move species from Appendix I to Appendix II. It is an extremely difficult and costly exercise to get a species or a national population of a species off Appendix I. The removal of the Australian population of the saltwater crocodile (*Crocodylus porosus*) from Appendix I, a move considered necessary for effective management, took a period of more than five years and an expenditure in excess of one million dollars. Only then was sufficient scientific data obtained to satisfy the Conference of the Parties that the proposal was justified.

The frequency of the transfers to Appendix I and the difficulty of reversing the decisions might not have mattered so much if Appendix I listings were themselves satisfactory from the point of view of conservation. Many think that they are not. Some exporting countries hold that Appendix I listings for certain species have had a profoundly negative impact on conservation. It is argued that CITES is only concerned with the threat to wildlife from international commercial trade and that there may be other threats. In some cases the decline in species is a consequence of conflicts with farmers. In these circumstances a prohibition on trade can actually be damaging to a species because it will remove most of the economic value of the species. This destroys the incentive to conserve wild populations in the face of economically important alternative land-uses.

A good example of the negative impact of an Appendix I listing is provided by the leopard (*Panthera pardus*). This was included in Appendix I in 1973 and, as a consequence, landholders came to see it not as an asset and something to conserve, but as a nuisance to be exterminated. The leopard had remained relatively common throughout many parts of its range in sub-Saharan Africa, because of its ability to prey on agricultural livestock. But the Appendix I listing removed its economic value to landholders and so it was perceived simply as a pest. It was shot and poisoned and as a result the populations of the species in some areas declined. So, even though the species was subject to the most stringent controls offered by CITES, these had very little effect on the ground where landholders (and governments) were faced with quite different management problems. In response to a concerted effort by many range states to reverse the decline and acquire the necessary flexibility to apply management strategies that conferred an economic value on the species, the fourth Conference of the Parties (Gaborone, 1983) instigated a system of national export quotas for the sub-Saharan population of leopards. This approach has proved very successful and has since been extended and applied to national populations of the cheetah (*Acinonyx jubatus*) at the ninth Conference of the Parties (Fort Lauderdale, 1994). The tenth Conference of the Parties (Harare, 1997) adopted a similar system for the markhor (*Capra falconeri*) populations in Pakistan.

Perhaps the best known protest against an Appendix I listing came from a number of southern African Parties to the convention who questioned the value of the transfer of the African elephant (*Loxodonta africana*) from Appendix II to Appendix I in 1989. Although undoubtedly warranted for some populations, this listing was seen by those countries, in which elephant numbers were stable or increasing, as an example of an unwarranted and unnecessary intervention by the international community. Thus, Appendix I listings are by no means universally regarded as the panacea for all conservation problems.

A final factor increasing the importance of the failure to implement Article IV effectively has been the greater willingness of some Parties to use stricter domestic measures where they fear that Appendix II species are being traded unsustainably. Many economically powerful importing Parties, perhaps reluctant to change the listing of a species, or frustrated with the length of time which the movement of a species to Appendix I from Appendix II can take, have settled on a second response which is based on Article XIV(1) of the convention. This states:

> *The provisions of the present Convention shall in no way affect the right of Parties to adopt (a) stricter domestic measures regarding the conditions for trade, taking, possession or transport of specimens of species included in Appendices I, II and III, or the complete prohibition thereof, or (b) domestic measures restricting or prohibiting trade, taking, possession or transport of species.*

Citing this article as a justification, there has been a rapid growth of cases where the scientific authorities of importing Parties take it upon themselves to determine if the trade in an Appendix II species is detrimental. They then issue their recommendations on whether those species can be imported. These determinations are often made without consulting the exporting country and often with incomplete or out-of-date information. Different Parties and groups of Parties have implemented these stricter domestic measures in different ways and with different degrees of consultation. But, in practice, the use of such unilateral measures, while often resulting in good conservation, has been seen as usurping the authority of the exporting countries. It has caused a great deal of resentment and tension within the convention, which is a multilateral agreement depending on the goodwill and mutual respect of the participating countries.

All of these considerations – the unsustainable trade in some Appendix II species, the difficulties of transferring species from Appendix I to Appendix II, the unsatisfactory nature of Appendix I itself, and the rise in the use of stricter domestic measures – created a strong incentive to improve the implementation of Article IV.

IMPROVING ARTICLE IV:
THE SIGNIFICANT TRADE PROCESS

It was against this background that the CITES Animals Committee, in its preparations for the eighth Conference of Parties (Kyoto, 1992), made the first serious attempt to address the problems related to the implementation of Article IV. Their aim was to prevent Appendix II species becoming further endangered. This would stem the rate at which many species, often without the support of the range states, were being transferred from Appendix II to Appendix I. Coincidentally, a related debate was occurring in another forum in the lead-up to the second United Nations Conference on Environment and Development that was to take place in Rio de Janeiro the same year. A series of Preparatory Committee meetings, attended by a large number of developing countries, provided the means for the 'resource rich' countries of the third world to formulate uniform negotiating positions for the forthcoming UNCED 2. The issues discussed there included: the sustainable use of wildlife; the access to genetic resources; and greater equity in benefit sharing. These issues were to become the pillars of the CBD and Agenda 21.

Resolution Conf 8.9 on Trade in Wild-caught Animal Specimens was adopted by the eighth COP (Kyoto, 1992) as a means to facilitate the improved implementation by exporting Parties of Article IV(2)(a) and (3). The resolution establishes what is usually known as the significant trade process. It was adopted in the recognition that, when an Appendix II species becomes eligible for inclusion in Appendix I of the convention, it is not just a sign that an exporting country has not implemented the requirements of Article IV; it is also a serious conservation failure and an indictment of the whole convention.

The significant trade process seeks, in cooperation with the Management Authorities of exporting countries, to identify and rectify Article IV implementation problems. The process entails an initial assessment of the available trade data for Appendix II-listed animals in order to determine those taxa which are possibly being traded in excessive quantities. Following agreement by the Animals Committee on the composition of candidate taxa, more detailed reviews, incorporating greater consideration of the biological characteristics of each species, are conducted by the World Conservation Monitoring Centre, IUCN and TRAFFIC-International under contract to the CITES Secretariat. Draft reviews are then circulated to the Management

Authorities of range states for comment, correction or updating, before being considered by the Animals Committee.

The significant trade process requires the CITES Animals Committee to assess all available information and determine whether or not the provisions of Article IV(2)(a) and (3) have been satisfied. In cases where a particular problem has been identified, the Committee formulates primary or secondary recommendations designed to correct the problem. These recommendations from the Animals Committee are communicated by the Secretariat to the Management Authority of the relevant exporting country. Problems of a serious nature become the focus of primary recommendations. In these cases the recipient Party has 90 days in which to respond to the satisfaction of the Secretariat. Less serious problems or problems requiring field studies become the subject of secondary recommendations for which the recipient country has twelve months to submit a satisfactory response to the Secretariat. In cases where the exporting Party has received a primary or secondary recommendation but either fails to respond to correspondence from the Secretariat, or provides an unsatisfactory response, the Secretariat is authorized to refer the matter to the CITES Standing Committee. In extreme cases, the Standing Committee has recommended that Parties refuse imports of a particular species from a country until such time as that country has addressed the problem to the satisfaction of the Secretariat. In many cases the Management Authority of the exporting country consults the Secretariat in order to define a more conservative annual export quota that is deemed to fulfil the requirements of Article IV(3).

The effective implementation of Article IV should ensure that we do not continue to watch species decline while they enjoy the supposed advantages of an Appendix II listing. If the appropriate corrective measures are taken at the appropriate time, then there should be a reduction in the number of animal species which need to be transferred from Appendix II to Appendix I. In this regard, it would be informative, when proposals to transfer taxa from Appendix II to Appendix I are being evaluated, to consider whether or not the taxon has been subject to review in terms of the significant trade process pursuant to Resolution Conf 8.9. If it has been subject to review it should then be established if and why the process has failed the species. Only in this way will we learn enough to modify and improve the process.

There are some weaknesses in Resolution Conf 8.9. One of these concerns funding. The most common recommendation arising from this resolution focuses on the need for field surveys for Appendix II species in order to arrive at scientifically based harvest and export quotas. But if the funding for these field surveys is not available, then they will not be undertaken. To date, only a small proportion of the developing countries that have been required to undertake fieldwork have been able to do so. As the Resolution Conf 8.9 process becomes more institutionalized, it will be essential also to 'institutionalize' a reliable, on-going source of funds to undertake field studies. Past contributions by various donor agencies and governments have enabled specific studies to be undertaken. But the availability of funds from these sources is not guaranteed and when they are made available they are often tied to a particular country or species. This may or may not fit with the priorities identified by the Animals Committee and the CITES Secretariat.

Another problem concerns the potential for abuse. Some protectionist and animal rights organizations perceive the Resolution Conference 8.9 process as a means of achieving their ideological goal of prohibiting international trade in wild animals. Indeed, some parties that have been the recipients of recommendations have expressed concern that the process represents a 'back-door' mechanism to achieving the same effect as an Appendix I listing. This perception is reinforced further if importing countries use the process to justify the application of unilateral import bans on species that have been subject to Resolution Conf 8.9 recommendations, irrespective of the exporting country in question taking the necessary corrective action to the satisfaction of the Secretariat. In light of these concerns, it is essential that all Parties understand and participate fully in the process. As a committee representing the interests of the Parties, it is also important that the CITES Animals Committee maintains an objective and cooperative approach to the implementation of Resolution Conf 8.9. It is not clear that sufficient mechanisms are in place to ensure that this is always the case. But it is a positive sign that meetings of the Animals Committee have shown near exponential growth in the attendance of both Parties and observers, including non-governmental organizations (NGOs). While some legitimate concerns have been expressed about the cost of this and the burden it places on developing countries, it seems a reasonable price to pay for an effective convention.

Despite the potential for the abuse of Resolution Conf 8.9, it does offer something to both sides of the philosophical divide which CITES straddles. On the one hand, protectionists have long argued that exporting countries must be obliged to make Article IV non-detriment findings using scientific methodology. The significant trade process proves a mechanism for reviewing the non-detriment findings of range states. On the other hand, the supporters of sustainable use can be pleased that the significant trade process provides a mechanism whereby trade can continue and the options for the management of wildlife remain flexible. The significant trade process is a compromise in which the Parties have acknowledged that CITES cannot operate as it is currently structured. They have surrendered a little of their sovereignty to ensure that the convention can achieve its objectives through a multilateral process that involves a high degree of consultation and cooperation. Although the significant trade process can result in punitive measures, such as specific trade sanctions being applied to a Party for failure to implement the provisions of Article IV, it is implicit in the process that the species remains on Appendix II and the exporting country retains ultimate control over the management of the species. Furthermore, the use of the significant trade process when there are problems with the implementation of Article IV, generally removes the need for importing countries to apply stricter domestic measures such as import bans or independently derived import quotas. This is a feature which is viewed very positively by exporting countries. Another positive feature associated with the significant trade process, and one which was probably important in ensuring that exporting countries were prepared to adopt the process in the first place, is the fact that it can result in individual exporting countries being given assistance to develop the necessary technical and administrative capacity to implement the requirements of Article IV. All in all, the significant trade process offers an effective way of patching over some of the cracks in the convention.

CONCLUDING REFLECTIONS

It is arguable that the significant trade process offers more than a means of remedying some of the weaknesses of CITES. It could provide the key to a radical overhaul of the relations between CITES and the CBD. Under the framework of the CBD a specific protocol could be developed for the control of the harvesting of wildlife both for internal purposes and for international trade. This protocol could treat the significant trade process, as introduced in Resolution Conf 8.9, as a model. Different harvesting and trade regimes could be applied to species depending on where they lie on a single 'conservation continuum' that ranged from highly endangered to extremely abundant. There would be no need for the binary system of Appendices I

and II. Trade could be subject to quotas or other management measures, or even halted completely, to ensure that recorded trade levels
do not impact adversely on populations of the species in the wild.
Most significantly, as an instrument within the framework of the
CBD, such a protocol would be eligible to receive funds through the
Global Environment Facility for its operation – and the necessary
field studies might then get funded. Perhaps it is not too late to move
in this direction?

Chapter 6

Who Knows Best?
Controversy over Unilateral Stricter
Domestic Measures

J M Hutton

*The provisions of the present Convention shall in no way
affect the right of Parties to adopt: (a) stricter domestic measures regarding the conditions for trade, taking, possession
or transport of specimens of species included in Appendices
I, II and III, or the complete prohibition thereof; or (b) domestic measures restricting or prohibiting trade, taking, possession or transport of species not included in Appendix I, II
or III.*

CITES Article XIV, Paragraph 1

INTRODUCTION

Every two years or so, over 140 countries meet at the Conference of
the Parties to CITES. The debate is vigorous and very often the Parties, being unable to find consensus, vote to establish the majority
position. In this way the Parties determine the appropriate appendix
listing for each species of interest and draft the resolutions and decisions of the convention. Together, the listings, resolutions and decisions constitute the detailed system that all member countries are
expected to follow until a subsequent Conference of the Parties de-
cides otherwise.

Unfortunately, not all Parties do follow the detailed system as
agreed by the majority. Some take advantage of Article XIV and apply 'stricter domestic measures'. Being, by definition, stricter than

CITES, these measures usually end up prohibiting trade that CITES has otherwise agreed is safe. Stricter domestic measures can take several forms. The majority are taken by producer states which choose to apply domestic controls on trade which are stricter than those of CITES, sometimes for conservation reasons. The most controversial are those taken unilaterally by economically powerful consumer nations. These move the judgment of sustainability from the producer state (for whom it is an obligation under Article IV) to the consumer state. They also run counter to the 'agreements' (the listings on the appendices) reached between producers and consumers at the meetings of the Conference of the Parties. While some systems for implementing stricter domestic measures create more opportunities for dialogue between the consumer and producer nations than others, many developing countries view any unilateral imposition of stricter domestic measures in the face of common agreement as an affront.

Such actions are particularly unpopular in southern Africa, much of which is only just reestablishing its jurisdiction over resources after a colonial past. This region has been outspoken in arguing that there is no room for unilateralism in a multilateral environmental agreement that already requires participants to surrender some of their sovereignty. It is contended that those countries which are creating impositions and seeking to judge the competency of others have a range of experiences that may not be relevant to countries of the developing world, and in any case they are themselves often severely lacking in resource management credentials. There is some substance to these claims. At the most simplistic level, it would be hard to argue that the wildlife conservation record of Australia, Germany, France or the US, for example, is better than that of Botswana, South Africa, Tanzania or Zimbabwe. However, it is perhaps more constructive to note that those countries with better developed industrial economies have, by and large, a history of wildlife depletion followed by a period of restoration, often involving strict protection. It should therefore be no surprise that their conservation policies are often well out of step with those of resource-rich developing countries.

Furthermore, all governments carry the responsibility for the wildlife within their borders and are accountable for its fate to their people. Stricter domestic measures that remove or undermine the authority for decision-making from the governments of developing countries leave them with the responsibility and accountability, but with fewer options for action. At the same time, those states which take the stricter domestic measures end up with the authority without the responsibility or accountability – which is a sure recipe for irresponsibility.

Under these conditions one might predict some short-term gains from stricter domestic measures, but also some serious conflicts between the conservation measures adopted by resource rich developing countries and those advocated by consumers. Arguably, this is what has happened. The benefits of stricter domestic measures are usually uncritically accepted (see below), but only rarely is it appreciated that there are examples where inappropriate measures taken by consumer countries have threatened the success of conservation programmes. The most numerous and best documented are to be found amongst the crocodilians. For instance, in 1983 the Nile crocodile population of Zimbabwe was transferred to CITES Appendix II to allow nearly unfettered trade in leather and leather products. This was done in recognition of the fact that there was a management programme in place that was leading to real conservation gains for the species. But despite the best efforts of the Zimbabwe Government and conservation groups, including the IUCN Species Survival Commission (SSC) Crocodile Specialist Group, and even in the absence of opposition, it took an additional 13 years for the US Government to allow commercial shipments of Nile crocodile leather and leather products to enter its markets. For other crocodile species, trade is still prohibited, irrespective of the position of CITES and any conservation benefits that such trade might have for the species.

Less negatively, it is worth noting that CITES Article XIV measures have not only been directed at particular species; they have also been used as a tool for ensuring general compliance with CITES. On occasion the Standing Committee, when faced with Parties which are not implementing their CITES obligations, has recommended that the broader CITES community impose sanctions, such as trade bans, on the errant Parties. These bans are essentially stricter domestic measures, but in this case they are not entirely unilateral and they are targeted at Parties delinquent in the implementation of CITES with whom (it is expected) consultation has been tried and failed. A good example of this sort of action are the trade sanctions imposed by the US against Taiwan in 1993, after the Standing Committee recommended such sanctions on the grounds that Taiwan had not taken sufficient action to eliminate the domestic trade in rhino horn. Many observers of CITES support this use of stricter domestic measures because the decision is made on a multilateral basis. However, since the Standing Committee last adopted this approach its composition has been changed to represent more equitably the geographic spread of the CITES membership. As a result, measures of this sort, which tend to be favoured by economically better off countries, are expected to be adopted less often in future.

Benefits and Assumptions

It is widely assumed that tangible conservation benefits flow from unilateral stricter domestic measures of the type that allow large consuming countries to appropriate authority from developing countries. For example, the 1996 report, *How to Improve the Effectiveness of CITES*, commissioned by the Standing Committee, while noting some concerns about stricter domestic measures, had this to say on the subject:

> *While such measures can be important to the conservation of various species, and while the adoption of such stricter measures is well recognized in international law, the application of this right has led to concerns over equity and raises questions over the compatibility of CITES with the GATT/ WTO.*

It continued in a similar vein:

> *the use of stricter domestic measures provides a means of conserving certain listed species from over-exploitation from international trade but can also cause confusion and ill-feeling when they override recommendations made by the Conference of the Parties.*

However, the report does not provide examples of stricter domestic measures which have been 'important to the conservation of various species', or of the way in which they provide 'a means of conserving certain listed species from over-exploitation'. Presumably, the authors of the report, in common with many others, considered these to be self-evident truths. However, there has been no systematic analysis of the impact of the unilateral decisions that major wildlife importers take in the name of stricter domestic measures. While it tends to be assumed that closing a market automatically contributes to the conservation of the species concerned, trade has, in fact, proved to be flexible. It can shift between species, producer and consumer states. Where demand continues in the absence of legal supply routes it can quickly go underground, causing the proportion of illegal trade to increase dramatically relative to the legal trade. In the absence of systematic studies, it is almost always unclear what effect stricter domestic measures by importing states actually have in practice. Some studies that touch on the subject do suggest that the restriction and prohibition of trade can be an important conservation tool, but others demonstrate that it is far from a universal panacea. In some cases it can be highly detrimental to pragmatic conservation efforts.

Despite the 1996 report commissioned by the Standing Committee, the issue of stricter domestic measures did not receive much attention at the tenth Conference of the Parties in Harare in 1997. The only relevant output was Decision 10.103, which stated that 'a survey of stricter domestic measures already adopted by the Parties shall be carried out and a report shall be submitted to the Standing Committee, which shall consider a second stage of review'. If the survey is restricted to the compilation of a simple list then the convention will make little progress. What is actually needed is a more comprehensive study of the way that stricter domestic measures have been applied, including a comparison of the approach taken by the US and European Union (EU), two of the principal consumers of wildlife products which routinely invoke these measures. We need to know what the conservation gains have been (if any) and what have been the costs, and to whom. Which has been most effective, the 'top-down' approach of the US, or the more consultative mechanism of the EU? What can be done to decrease the frequency and improve the quality of stricter domestic measures? At the time of writing, the EU has commissioned a review of the stricter domestic measures that it has taken, but the US does not appear to be making any progress in this direction and coordination seems unlikely in any event.

STRICTER DOMESTIC MEASURES AND THE WORLD TRADE ORGANIZATION

The whole purpose of CITES is to achieve the protection of particular species through trade regulation including, where necessary, the total prohibition of trade whose purpose is 'primarily commercial'. Not surprisingly, given the value of trade in wild species, there is serious debate about the compatibility of CITES with the General Agreement on Tariffs and Trade (GATT).

There are several provisions of CITES which might be considered GATT unfriendly. For example, Article 1 of GATT requires World

Trade Organization (WTO) members to treat 'like' products in the same way irrespective of their origin or method of production. This would appear to conflict with split-listing within CITES. Split-listing has the consequence that some countries are allowed to trade in threatened or endangered species while others are prohibited from doing so. For various reasons, including the fact that it would chal- lenge the integrity of a range of multilateral environmental agree- ments, it is highly unlikely that there will be a CITES-related WTO dispute in this area. However, the same cannot be said for stricter domestic measures, which appear to be fertile ground for a dispute within the framework of WTO.

Here it is illustrative to return to the example of the Nile croco- dile and the way in which the US, through its Endangered Species Act (ESA), prohibited the entry of crocodile products for more than 13 years after CITES had agreed that trade was permissible. This case had several features that made it an ideal candidate for a challenge within the WTO. First, the US had its own 'like' product, alligator leather, which was in commercial competition with Zimbabwe's crocodile leather. Furthermore, the US case was weakened by the fact that the move to lift the prohibition under the ESA was unop- posed, even by the animal rights lobby (which tends to complicate matters where the ESA is concerned). In addition, some five years before any change was forthcoming, the directorate of the relevant US authority had gone on record as saying that there was no conserva- tion basis for the continuation of the ban on Nile crocodile products from Zimbabwe. However, although it had some special features, this case, in which stricter domestic measures were taken to restrict trade in an entirely arbitrary manner, was far from unique. Given that the dispute mechanism within CITES has restricted scope and is untried, the WTO dispute mechanism may well find itself the final arbiter of the legitimacy of some stricter domestic measures.

Fundamentalism vs Evolution

If stricter domestic measures are an affront to developing countries, as well as GATT unfriendly, while at the same time their effective- ness is unmeasured, why are they so enthusiastically adopted by many of the main wildlife markets of the world?

The obvious answer hinges on the non-detriment requirement of Article IV. In the way that CITES is structured, Appendix II listings have little impact unless exporting nations take seriously their obligation to ensure that trade is non-detrimental. The conspicuous failure of some developing countries to satisfy the non-detriment requirement has therefore provided an incentive for consumer nations to attempt to solve the problem by the adoption of stricter domestic measures in which they make the judgment of sustainability. However, an examination of the way in which Article XIV was introduced to the convention together with the way that two of the main consumers, the US and the EU, have subsequently implemented stricter domestic measures, suggests that there are some fundamental philosophical issues to be teased out in any meaningful discussion of unilateral action in CITES.

The idea that Parties might be allowed to adopt measures stricter than CITES appears to have been introduced to CITES during the negotiating phase. It follows the precedent that had been established in contemporaneous US law, notably the Marine Mammal Protection Act of 1972 and the Endangered Species Act of 1973. Both laws removed jurisdiction with respect to marine mammals and endangered species from the constituent states of the US and centralized it in the federal government. This 'pre-emption', as it was called, said that a state had to have laws at least as strict as the federal laws in order to recover the jurisdiction, but made it clear that nothing would prevent a state from having a 'stricter' law. The word 'stricter' was not defined, but the assumption behind this legislation was that wildlife use is bad and the commercial use of wildlife is even worse. Thus, 'stricter measures' was interpreted to mean a law that prohibited even more types of use than the already restrictive federal legislation. The way the US implements stricter domestic measures within CITES is no different. It commonly imposes trade embargoes and prohibitions on the use of species already listed by CITES, even where the international community has determined that the species' conservation might benefit from controlled trade. On 'precautionary grounds' the US has applied to threatened species the same stringent prohibitions that are required only for endangered species. Many of the species it regards as threatened are also listed on CITES Appendix II, so the result is a series of domestic regulations which are stricter than CITES with little or no technical justification. The twin elements of US action within CITES therefore remain the same as within its domestic laws – the assumption that harvesting and trade are inherently incompatible with conservation and the view that it is legitimate for constituent states to take stricter measures than the 'higher' body.

There are similarities between the stricter domestic measures taken by the EU and the US. They both take it upon themselves to make non-detriment findings for the range states and, in doing so, they both assume that they know best how developing countries should manage their resources. Nevertheless, the EU actions are not so clearly underpinned by the fundamental belief that wildlife trade and conservation are mutually exclusive. Instead, the main motivation of the member states of the EU seems to be the belief that the enforcement of sustainability is best moved from the producer to the consumer state because economic theory suggests that developing countries often have strong economic incentives to 'mine' resources rather than manage a 'production process' which results in sustainable resource management. This is reflected in the type of precaution inherent in the EU process; it imposes conditionality with respect to trade rather than inflexible trade bans.

Both the old (1984) and the new EU Regulations (applicable from 1 June 1997) go further than CITES by requiring an import permit for all Appendix II species. This can only be issued when the member states of the Union are convinced that current or expected levels of trade will not have a harmful effect on the population of the species in the country of origin. Under the old regulations the EU cancelled all trade in a particular Appendix II species unless the exporting state met its CITES Article IV obligations to the satisfaction of the EU member states. On many occasions the EU decided that there were shortcomings in the management of the wildlife trade and consequently refused to issue import permits. Subsequently, the EU member states would negotiate, on a country by country basis, the terms and conditions upon which they would allow imports. This often involved the negotiation of country-specific quotas enforced by the EU customs inspectors. Under the new regulation, the system is similar. If the EU member states are unable to make a 'non-detriment' finding they decline to issue import permits and the European Commission begins a process to formalize the import suspension. This involves obligatory consultation with the range states concerned and because the import restrictions are temporary in nature, the restrictions can be removed if new information is received and circumstances are considered to have changed.

The net result of the philosophical differences underlying the stricter domestic measures taken by the US and the EU is that the EU CITES regulation, as it is currently administered, has an overarching consultative element and a flexible mechanism for adjustment which allows the negotiation of 'conditionality' often to the satisfaction of both the EU and the range state concerned. The US system, in contrast, has a weak consultative process. The ESA only requires that the US try to notify foreign governments and to take into account any efforts they may be making to protect the species. The limits of budget and time have usually meant that the consultation that should occur on such important matters has not, in fact, taken place. Even if consultation does take place, the ESA does not contain the flexibility for positive conditionality to evolve.

THE ALTERNATIVE TO UNILATERALISM

Timothy Swanson argues in Chapter 12 that the evolution of CITES into a constructive trade control mechanism must involve the consolidation and globalization of the EU regime of conditionality to include all of the major consumer states. Otherwise, the imposition of stricter domestic measures by one consumer will often shift trade to others. This proposal is not attractive to the developing world. But if, as it seems, this is the route that CITES is to follow, then there are compelling reasons why this 'consolidated conditionality' should at least be achieved in a consultative and multilateral manner. Fortunately there are some indications that this will be the case.

Although the failure of developing countries (in particular, but not exclusively) to implement Article IV effectively is the main justification for the stricter domestic measures taken by key consumers, improving the way that non-detriment findings are made has not proved easy. In the most acute cases, where the same Parties are implicated time and time again, a large part of the solution must lie in the transfer of expertise and resources to the exporting countries. These often need assistance to overcome a lack of sound wildlife trade and conservation policies, limited scientific know-how, poor management capacity or an inadequate legal infrastructure. However, because this is a long-term strategy, which is expensive and may not always be successful, the Parties to CITES have explored another avenue – one that essentially acts as a safety net for the convention. This is known as the significant trade process. It operates very much like the EU model of stricter domestic measures, but on a multilateral and consultative basis (see Chapter 5). If this is effectively managed and implemented then unilateral steps taken by individual Parties or groups of Parties will be redundant – or certainly much harder to justify. The Parties to CITES would be well advised to amplify this process.

With the development of the significant trade process it is to be hoped that consumer countries will not take stricter domestic measures which prohibit trade unless they have exhausted all of the convention's mechanisms. Those taking stricter domestic measures should be expected not only to demonstrate that the CITES listing is inappropriate, and that legal commercial trade continues to have a negative effect on the species despite the CITES listing and the operation of the significant trade process. They should also be expected to show that the measure they are taking has been discussed with the range state concerned and can reasonably be expected to have a positive conservation effect. Finally, the system would be greatly facilitated if some economically dominant Parties within CITES abandoned the philosophically driven presumption that commercial wildlife trade is inherently contrary to conservation.

Part III

CASE STUDIES

Chapter 7

Assessing CITES: Four Case Studies

Michael 't Sas-Rolfes

INTRODUCTION

There are various factors that threaten the conservation status of wild species of fauna and flora. Among these, habitat conversion, fragmentation and destruction account for the most species losses. Excessive commercial exploitation accounts for a much smaller, but still significant, proportion of losses. CITES is intended to protect those species that are threatened by excessive commercial exploitation. To do this, it focuses on a very narrow aspect of commercial exploitation, namely transactions that take place across international borders (ie, international trade). CITES is not designed to address issues such as supply mechanisms, domestic trading regimes or consumer demand. CITES is, therefore, very limited in its potential effectiveness as a conservation tool. Not only does it fail to address issues of habitat loss, but it also fails to create mechanisms to control the supply of wildlife products and it has no direct means to influence consumer demand. As currently structured, CITES operates as a largely restrictive mechanism rather than as an enabling one. Implicit in its existing structure is an assumption that all trade is somehow bad for conservation unless proven otherwise. CITES measures, therefore, tend to emphasize limitations on trade rather than ways to facilitate trade that may ultimately enhance the status of wild species.

In theory, CITES is supposed to supplement, not replace, effective control of the supply of wild species (field protection). In practice, however, there are many cases where field protection is completely lacking and CITES provides the only readily available mechanism for controlling commercial exploitation.

Can CITES trade measures replace the need for effective field protection? The following four case studies suggest that it cannot. Each of these cases highlights serious shortcomings of the existing CITES mechanism and they offer some insights that can be used to design more effective wildlife trade regulation policies and mechanisms.

CASE STUDY 1: RHINOS

Background

There are five extant rhino species, two in Africa and three in Asia. The African species are the black rhino and the white rhino. Black rhino numbers have dropped from an estimated 65,000 in 1970 to about 2,600 in 1998. In the last six years numbers have increased in three range states: South Africa, Namibia and Kenya. Elsewhere, they continue to decline. There are two separate populations of white rhino. The northern population declined from some 2,000 in 1970 to a single population in Zaire of 17 in 1984. Since then this population has increased to about 25 and remains highly endangered. The southern white rhino was almost extinct at the turn of the century and was reduced to a single population of perhaps 20 animals in the Hluhluwe–Umfolozi district in South Africa. However, with careful management numbers have grown to some 8,440 today, and continue to increase.

The Asian species are the Indian, Javan and Sumatran rhinos. Indian rhino numbers have fluctuated; there was an increase in the early 1980s followed by a decline in the late 1980s and another recovery in the 1990s to a level of 2,041 animals by 1998. Javan rhino numbers appear to have remained fairly stable over the last decade or two, at some 70 animals. Sumatran rhino numbers have dropped considerably during the 1990s from an estimated 600–1,000 to the existing level of about 400.

In Africa, black and white rhinos were widely exterminated by hunting until conservation measures were implemented. Subsequently, rhinos have been eliminated by poaching for their horn. In Asia, the forest-dwelling Javan and Sumatran rhinos have been largely eliminated through habitat loss, although poaching for rhino products has also played a role. Indian rhinos have been affected by habitat loss, hunting and poaching for horn.

Rhino horn is a highly sought-after commodity. The horn of both Asian and African species is used as an ingredient in traditional Chinese medicines, to treat serious fevers and various other ailments. African horn is also used in Yemen to carve traditional dagger handles.

Other rhino body parts are also used in traditional medicines, especially in Southeast Asia where virtually every single body part has some use.

Ironically, the southern white rhino was probably the rarest of all rhino species and subspecies at the turn of the century, whereas today it is more numerous than all the other rhino species put together. The southern white rhino is the only true rhino conservation success story and in examining CITES' policies towards rhinos it is worth considering the factors that have contributed to this success and contrasting these with the factors that have led to the decline of all other rhinos.

CITES measures

The white rhino and the three Asian species were listed on CITES Appendix I at the founding conference in Washington DC in 1973. The black rhino was moved to Appendix I in 1977. After the Appendix I listings, the price of rhino horn rose dramatically in all consumer markets. For example, in Japan, recorded import prices per kg increased from US$75 in 1976 to US$308 in 1978; in South Korea prices increased from US$49 in 1976 to US$355 in 1979 and US$530 in 1981; and in Taiwan they rose from US$17 in 1977 to US$477 in 1980. In Yemen, the wholesale price of horn increased from US$764 in 1980 to US$1,159 in 1985. Trade continued despite the ban and demand was further fuelled by speculative stockpiling ('t Sas-Rolfes, 1995).

The Appendix I listings of all rhino species had no discernible positive effect on rhino numbers, and did not seem to stop the trade in rhino horn. If anything, the Appendix I listings led to a sharp increase in the black market price of rhino horn, which simply fueled further poaching and encouraged speculative stockpiling of horn. Recognizing the failure of the Appendix I listing, in 1981 the Parties at the third Conference of the Parties passed a resolution (Resolution Conf 3.11) on the rhino horn trade. This resolution called on nations that were not Parties to CITES to also take measures to prevent the international trade in rhino products, and it called for a moratorium on the sale of all government and parastatal stocks of rhino products. Subsequent to this resolution, rhino poaching and trade continued unabated in most African countries. For example, between 1981 and 1987 Tanzania's black rhino population dropped from 3,795 to about 275 and Zambia's dropped from 3,000 to just over 100 (Milliken, Novell and Thomsen, 1993).

The obvious failure of Resolution Conf 3.11 prompted a further resolution to be passed at the sixth Conference of the Parties (COP) in 1987. This resolution (Resolution Conf 6.10) called for even stricter measures, including the complete prohibition of trade in all rhino products, both internationally and domestically. It also called for the destruction of government stocks of rhino horn, and suggested that affected countries should be financially compensated for destroying their stockpiles. Since the 1981 Resolution had been ignored by the governments of several countries the new resolution recommended that Parties should exert political, economic and diplomatic pressure on any countries that 'continued to allow the trade in rhino horn'. This later resolution was again ignored by several consumer countries and range states. Most range states refused to destroy their stockpiles of rhino horn and several key consumer countries failed to implement domestic legislation. Rhino horn trade and poaching continued: for example, Zimbabwe's black rhino population was reduced from 1,750 animals in 1987 to 430 in 1992, despite a policy of shooting poachers on sight. To protect its remaining rhinos, the Zimbabwean Wildlife Department had them all dehorned and moved to a few intensive protection zones, where they remain under constant surveillance by heavily armed guards.

Dissatisfied with the performance of the CITES ban, the governments of South Africa and Zimbabwe concluded that it would make more sense to allow a controlled legal trade in rhino horn. Wildlife departments in both countries had obtained significant stockpiles of horn through seizures from illegal traders, dehorning operations and the retrieval of horns from dead animals. At the eighth COP in 1992 South Africa submitted a proposal to downlist its white rhino population to Appendix II and Zimbabwe did the same for both its white and black rhino populations. These proposals were all rejected by the conference.

In 1992 the United Nations Environment Programme (UNEP) appointed a 'special envoy for rhinos', and provided him with funding to visit various countries to persuade their governments to abide by CITES. At the same time, the US government threatened four consumer nations with trade sanctions under the Pelly Amendment. This piece of US legislation empowers the US President to suspend all wildlife and fisheries trade between the US and any country considered responsible for diminishing the effectiveness of an international treaty designed to protect a threatened or endangered species. Governments of consumer nations responded to these pressures by passing laws and intensifying efforts to control illegal trade, but these efforts only served to drive the trade further underground.

In 1993 UNEP held a meeting in Nairobi to raise funds for rhino conservation. At the meeting, the range states requested US$60 million in emergency funds over the next three years, but only US$5 million were pledged over the next 12 months. South Africa reiterated its belief that a legal trade in rhino horn offered a potential solution, because sales of legally held rhino horn stockpiles could provide a substantial source of revenue to conservation agencies.

In South Africa, the Natal Parks Board ably demonstrated how effective commercial use and management could enhance the status of rhinos. After initially reintroducing white rhinos to many state parks and reserves, the Natal Parks Board embarked on a programme to re-establish white rhino populations on private land. White rhinos became increasingly popular among private landowners as a 'draw-card' species, both for trophy hunting and non-consumptive tourism (ie, for game-viewing purposes). Since 1986, the Natal Parks Board has auctioned white rhinos to the private sector. In 1990, the Natal Parks Board also starting auctioning black rhinos. Increased demand and rising prices for live rhinos have ensured that private landowners have a strong incentive to conserve and breed rhino populations. At the time of writing, at least 20 per cent of the white rhino population in South Africa is in private hands. Tourist viewing and trophy hunting revenues have been considerable and have mostly been reinvested in rhino conservation. The Natal Parks Board has also raised considerable revenues from its auctions, the proceeds of which are also reinvested directly into conservation.

After an initial meteoric rise, the prices of live white rhinos started to stabilize in the early 1990s. In 1994, at the ninth COP, South Africa again applied to have its white rhino population downlisted to Appendix II, subject to an annotation. The annotation provided that only live animals and trophies would be traded commercially. All other trade would continue to be prohibited. To the surprise of some this proposal passed comfortably. What effect did this have? At the subsequent 1995 Natal Parks Board auction, the average price of a live white rhino once again increased. This was certainly because the market for live white rhinos had been expanded to allow international bidders to participate in the auction. The outcome of the Appendix II downlisting was thus positive for conservation, as the Natal Parks Board was able to generate further revenues. At the tenth COP in 1997, South Africa applied to CITES to change the Appendix II annotation to allow for trade in parts and derivatives, but with a zero quota. This proposal was not accepted by the required majority of parties.

Lessons

The CITES Appendix I listing of all rhino species failed to stop either trade or poaching. With the exception of the southern white rhino, all rhino species appear to remain critically threatened. Although poaching levels have dropped in recent years and some populations appear to be increasing, it is not clear that this is as a direct consequence of the successful implementation and enforcement of CITES. Where there have been successful rhino conservation efforts, it appears to have more to do with high levels of field protection than with the successful implementation of CITES policies. So, if consumer demand for rhino horn should rise again in the future, the consequences for wild rhino populations could be dire. Conservation agencies would need to find increased funding for field protection at a time when the budgets of most conservation agencies are being reduced. This raises the question of where the increased funding is to come from.

The South African experience with the southern white rhino suggests a possible way forward. There, white rhinos have provided a source of revenue for their owners, and this has provided the incentive and the means to invest in rhino conservation. The next step may be to allow South Africa to sell legal stockpiles of horn. This is what the Natal Parks Board has been investigating. Unfortunately, however, the CITES system is steeped in politics. All South Africa's proposals to resume trade were rejected, either because there were no immediate or direct benefits for other range states with limited field protection measures in place, or because environmental lobby groups were concerned with the effect on more contentious proposals, such as the reestablishment of a legal ivory trade.

CASE STUDY 2: ELEPHANTS

Background

There are two extant elephant species, the African elephant and the Asian elephant. The African elephant definitely survives in nineteen range states and possibly in another eighteen. The Asian elephant survives in thirteen range states. African elephant numbers are thought to have dropped from more than 1.3 million in 1979 to roughly 632,000 in 1989, and were thought to be between 286,000 and 580,000 in 1995. Scientists estimated the Asian elephant population at between 30,000 and 55,000 animals in 1990.

The main cause for the African elephant's decline has been poaching for ivory. In contrast, the main cause for the Asian elephant's decline has been habitat loss and encroaching human population. Habitat loss and human encroachment is also a factor affecting the African elephant in some parts of its range and will become increasingly important in the future. Only fully mature Asian elephant bulls have tusks sufficiently large to be attractive to poachers; ivory poaching constitutes a lesser, but still significant, threat to Asian elephants. Both African and Asian elephants are also poached for their meat and hide in some parts of their range.

Traditionally, elephant ivory has been widely used for ornamental purposes. The demand for ivory is strongly entrenched in Asian culture. In Japan, ivory is especially prized for making traditional personal seals called hankos. Japan, Hong Kong and Singapore have been major centres for working ivory to make ornaments. Although previously substantial, the demand for worked ivory and ivory ornaments has dropped considerably in Europe and North America since the 1989 ivory ban. During the 1980s demand for ivory increased strongly in Asian countries such as South Korea and Taiwan (Barbier, Burgess, Swanson and Pearce, 1990) and there is evidence that this demand persists.

CITES Measures

The Asian elephant was listed on Appendix I at CITES inception. The African elephant was initially listed on CITES Appendix II, in 1976. This listing clearly failed both as a trade measure and as a conservation measure. In an attempt to make the listing more effective, special resolutions were passed at the third, fourth, fifth and sixth Conferences of the Parties.

At the fifth COP the Parties introduced a management quota system which took effect in 1986. A subsequent study by the Ivory Trade Review Group (ITRG) revealed that neither the management quota system, nor any of the earlier CITES resolutions were sufficient to control illegal poaching and trade. They concluded that the CITES Appendix II listing of the African elephant had been a failure. 'Weak management and enforcement capacity' was cited as the key reason for this failure.

The release of the ITRG report led to calls by major western environmental groups for a complete ban on the international trade in ivory. Thus at the seventh COP in 1989 the majority of the CITES Parties voted to list the African elephant on Appendix I. Eight African range states opposed this listing, arguing that they had adequate capacity to regulate illegal trade, but their protests were disregarded. Most proponents of the ivory ban favoured a blanket ban, to avoid possible laundering of ivory products through other countries. This was a valid concern, as countries such as Burundi and South Africa were shown to be major entrepôts for ivory poached in neighbouring range states.

The CITES Appendix I listing was accompanied by considerable media coverage and there was much anti-ivory publicity. Traditional western consumer markets in North America and Europe were noticeably affected by this and the demand for worked ivory products in these markets effectively collapsed. This had an obvious effect on the market price of ivory, which also fell substantially. Nonetheless, subsequent to the ban, there was evidence of ongoing demand, especially in Asian consumer countries. Ongoing demand for ivory is indicated by:

- continued poaching and illegal trade in certain African range states, especially those with poor levels of field protection;
- rising demand for substitute products such as hippo ivory; and
- records of mammoth ivory mining in northern Siberia.

At the two COPs subsequent to 1989, some African countries attempted to have their elephant populations downlisted back to Appendix II. For example, at both the eighth and ninth COPs South Africa submitted proposals to downlist its elephant population. At the eighth COP in 1992 it requested permission to trade in both ivory and hides, but this proposal was rejected. At the ninth COP in 1994 South Africa requested permission to trade in elephant hides only but after a show of vigorous opposition withdrew this proposal.

Shortly after the ninth COP a group of respected elephant scientists released a report arguing that the effects of the CITES Appendix I listing were mixed, with some range states reporting increased incidents of poaching. The report also noted with concern that field enforcement budgets were falling in most range states. The release of this report provoked an indignant reaction from many environmental groups, who seem determined to maintain the orthodox belief that the CITES ivory ban has been an unqualified conservation success.

At the tenth COP, held in Harare in 1997, Botswana, Namibia and Zimbabwe finally succeeded in having their elephant populations transferred to Appendix II. This transfer was only endorsed after a lengthy initial debate, the formation of a working group at the conference and extensive negotiations behind the scenes. The eventual decision, while it allowed these three countries to sell ivory again, was hedged around with numerous restrictions. Only specified amounts of ivory from the existing stockpiles could be sold, and only to one customer, Japan. Moreover, the sale could only begin 18 months after the transfer to Appendix II took place and the CITES Standing Committee was empowered to retransfer the populations to Appendix I if there was an escalation of illegal trade or hunting due to this reopening of the legal trade. Despite all these restrictions the three southern African countries greeted the decision as a victory for the principle of sustainable use.

At the time of writing, the CITES Standing Committee has granted approval for Namibia and Zimbabwe to sell certain stockpiles to Japan, and a decision on Botswana is still pending.

Lessons

The case of the African elephant has presented CITES with an interesting challenge. In certain African range states elephants are thriving and even becoming a potential problem. For example, in countries such as Kenya and Zimbabwe, expanding elephant populations are increasingly encroaching on areas inhabited by peasant farmers. These animals are a menace to the local people, destroying their crops and threatening their lives. In South Africa, elephant populations are confined within fenced protected areas, and their numbers may need to be controlled to prevent the adverse ecological impacts of excessive population pressure.

There are three ways to control elephant numbers. The first is through managed culling operations; the second is to translocate live animals to new areas, and the third is to use some form of elephant contraception. All three methods are costly and problematic. Elephant culling is costly and only profitable if the products from culled animals (meat, hides and ivory) can be sold at reasonable market prices. Elephant translocation, while more humane, involves considerable costs and is only a viable option as long as sufficiently large unpopulated areas of elephant habitat remain. Elephant contraception, a technique that is still being developed, also involves high costs, and raises ethical concerns (albeit different from those associated with culling).

It is ironic that scientists are grappling with ways to control expanding elephant populations when the rationale for the 1989 Appendix I listing was the imminent 'threat of extinction of the African elephant'. It is also ironic that African range states now possess in excess of 500 tonnes of stockpiled ivory worth millions of dollars, which they are mostly unable to sell, while their conservation departments are desperate for funds for field protection. It appears that the only sensible long-term approach is to re-establish some degree of controlled legal trade. Otherwise, the situation will simply get worse over time; ivory stockpiles will increase further and elephant populations will expand further in areas where they are well protected, thereby creating problems of over-population. At the same time, elephants will continue to be poached heavily in other areas where they are unprotected, thus providing further impetus for calls to maintain a complete trade ban.

The case of the African elephant demonstrates clearly that CITES Appendix II listings are ineffective in developing countries with neither the will nor the resources to implement the CITES system. The Appendix I ban appeared to work because of the fall in demand that resulted from the media publicity surrounding the ivory ban. However, some substantial markets remained, especially in East Asia, and these provided an incentive for continued illegal activity. In the longer term, a complete Appendix I listing of the African elephant is probably neither sustainable nor appropriate as a conservation measure, since it fails to address the ongoing demand for ivory products and the opportunity costs of conserving elephants in African range states. It remains to be seen whether a return to limited trade under a 'split-listing' is an effective solution, but there do not appear to be any better options.

CASE STUDY 3: TIGERS

Background

There are five extant tiger subspecies. These are the Bengal, Indochinese, Sumatran, Amur (Siberian) and South China tigers. A further three subspecies have already become extinct this century: the Caspian, Javan and Balinese tigers. At the start of the twentieth century,

wild tigers were widely distributed throughout Asia, with an estimated total population of 100,000 animals. Their range extended as
far west as Turkey, as far north and east as south-eastern Russia, and
as far south as the Indonesian islands of Java and Bali. Today the wild
tiger's range has been reduced considerably, but populations still survive in fourteen different Asian range states. According to recent estimates there are between 5,000 and 7,500 surviving tigers. Of these,
some 2,500 to 3,750 animals survive in India, making it the most significant range state. Tigers breed easily in captivity and there may be
as many animals in zoos and circuses as there are in the wild. Most captive animals are of mixed or uncertain pedigree but some 1,200 are
recorded as pure-bred specimens representing one of the five subspecies. There are concerted international efforts to establish healthy
captive populations of all five subspecies.

Tigers are threatened by loss of habitat, conflict with humans and
poaching. Habitat loss and fragmentation (including declines in natural prey species) appears to be the most significant reason for the tiger's decline. In areas where tiger conservation efforts are successful,
expanding populations tend to come into conflict with people, at-
tacking them and their livestock. Dead tigers have considerable commercial value; their skins and bones are especially prized. Tiger bone
is an ingredient in traditional Chinese medicines used to treat rheu-
matism and a wide range of other ailments. Most tiger populations
are poorly protected in the field, and many survive outside protected
areas, where they are especially susceptible to habitat destruction.

Until the early 1970s tigers were widely eliminated throughout
Asia due to habitat loss and hunting. Being largely regarded as pests,
tigers were frequently eradicated under government-sponsored 'bounty'
programmes. However, in the early 1970s, conservationists from India
called for measures to protect the rapidly disappearing Bengal tiger. In
1973 the Indian Government launched 'Project Tiger', an initiative
which was backed by a World Wide Fund for Nature (WWF) fund-
raising campaign called 'Operation Tiger'. The aim of Project Tiger
was to create a number of dedicated tiger reserves where tigers, their
habitat and prey would be protected. WWF also supported projects
outside India: in Nepal, Bangladesh, Thailand and Indonesia. Many
Asian countries passed stronger wildlife protection laws, banning tiger
hunting and creating new protected areas. These measures seemed to
be effective for the first 10–15 years; reports from India, Nepal and
the Soviet Union indicated that tiger numbers were once again increasing in those countries. Unfortunately this success was short-
lived.

CITES Measures

From the inception of CITES, all tiger subspecies were listed on Appendix I, except the Amur tiger, which was listed on Appendix II. In 1987 the Amur tiger was moved up to Appendix I. Subsequent to 1987 there were mounting concerns that the tiger's status was not as secure as previously thought. During 1987 the Chinese National Pharmaceutical Bureau had asked the Beijing Pharmaceutical Company to draw up plans for a tiger breeding facility near Beijing to 'solve the problem of the shortage of tiger bone' needed for the manufacture of medicinal liquor. There were clear indications that the prices of tiger body parts were rising. In 1992 serious levels of poaching were recorded for the first time in two high profile areas: India's Ranthambore Tiger Reserve and the Russian far east. This spurred the international conservation community into action, to address this perceived new threat of commercial demand for tiger parts.

The trade in tiger bone is especially hard to control because of the so-called 'look-alike' problem. Tiger bones are difficult to distinguish from the bones of other animals; in particular they are virtually indistinguishable from the bones of similarly sized felids such as lions. This fact, coupled with poor border controls between various Asian range states, compromises proper implementation of the CITES Appendix I listing. In an attempt to encourage greater enforcement from within consumer states, environmental groups persuaded the US government to threaten certain range states under the Pelly Amendment. This prompted some consumer countries to introduce enforcement measures, but it failed to end illegal trade and poaching.

At the ninth COP, in 1994, the CITES Parties noted that the range states had undertaken an initiative called the Global Tiger Forum to launch a worldwide campaign to save the tiger. The Global Tiger Forum developed slowly and does not appear to have achieved much at the time of writing. At the tenth COP the Parties passed a new resolution, which called for the Standing Committee to review tiger trade issues during the intersessional period (ie, before the eleventh COP) and to identify countries which need to employ additional measures to minimize illegal trade. In addition, the Standing Committee was asked to send technical and political assistance to such countries. 1998 was the Year of the Tiger in the Chinese calendar, and there was much NGO activity related to tiger conservation, including concerted campaigns to highlight trade issues. This is ongoing at the time of writing.

Lessons

A CITES Appendix I listing is an incomplete and possibly ineffective solution to the tiger conservation problem. It is incomplete, because the chief threat to wild tigers is not excessive exploitation, but rather loss of habitat and suitable prey species. As a trade measure, CITES does nothing to address this problem.

It also appears to have had limited success at influencing consumer demand or preventing illegal trade. This is partly because several consumer states also happen to be range states with much trade taking place domestically. CITES, of course, has no mandate to regulate domestic trade issues.

A possible alternative solution to excessive commercial exploitation of wild tigers is to provide an alternative supply of tiger products that is cheaper, legal and authenticated, thereby undermining the incentives for illegal exploitation and trade. This could be achieved by tiger farming, an option which is of interest to the Chinese government. But CITES is not well structured to facilitate the development of this option (which is also vigorously opposed by many environmentalists). And even if tiger farming were allowed under a registered captive breeding programme, there is no mechanism to ensure that proceeds from the sale of farmed tiger products would in any way be invested toward the conservation of wild tigers.

CASE STUDY 4: BEARS

Background

There are eight extant bear species. They include the American black bear, polar bear, brown bear, Asiatic black bear, sun bear, sloth bear and spectacled bear. These seven species all produce a substance called ursodeoxycholic acid (UDCA), which has unique medicinal qualities. The eighth bear species, the giant panda, does not produce UDCA, and some biologists dispute whether it is a true bear.

The American black bear is the most common species with as many as 800,000 individuals surviving on the North American continent. The brown bear is the second most numerous species (*circa* 180,000) and is widespread from Europe, through Asia, to North America. There are thought to be between 20,000 and 30,000 polar bears surviving in the arctic regions of North America, Europe and

Asia. The spectacled bear (*circa* 10,000) is found in the Andes region of South America. The remaining species, the Asiatic black bear (*circa* 50,000), sun bear (<50,000) and sloth bear (<10,000) all survive in parts of Asia. Populations of the American black bear and polar bear appear to be stable and even increasing. The brown bear is secure in some areas and declining in others, and numbers of the remaining four species continue to decline.

Worldwide, bears are threatened by habitat loss, bear–human conflict and excessive levels of commercial harvesting. Various bear body parts have commercial value, including gall bladders, paws, meat, skin, teeth and claws. The bile contained in bear gall bladders is highly sought after by Korean, Chinese and Japanese people who use it as a traditional medicine and health tonic. Bear meat, especially the meat from bear paws, is a prized delicacy and also considered to have therapeutic effects, especially by Koreans. To satisfy the demand for gall bladders and paws in Asia, bears have been harvested unsustainably in several Asian range states. For example, the Asiatic black bear may now be extinct in South Korea.

Most Asian range states have passed laws restricting the harvesting of wild bears, but the consumption of bear products is still widely accepted in countries such as China and Japan. Since the mid-1980s the Chinese government has encouraged bear farming. Captive bears are confined to small cages and have catheters surgically inserted into their gall bladders to enable bile 'milking'. In 1996 there were more than 7,500 captive bears on farms in China. Most of these animals are subjected to conditions that are considered unacceptable by animal welfare groups.

Wild bears are legally hunted in several countries including the US, Canada, Russia and Japan. Around 40,000 bears are killed every year in North America by trophy hunters alone; this is considered to be a sustainable level. Russia and Japan allow the harvesting of bear parts from legally hunted animals, as do certain jurisdictions within the US and Canada. Conservationists allege that there are high levels of illegal hunting and trapping in addition to the legal offtake of animals.

CITES Measures

Apart from Russia's brown and polar bear populations, all Asian bears have been listed on CITES Appendix I, as has the spectacled bear. The brown bear is 'split-listed', on Appendix I in some range states, and on Appendix II in others. The polar bear is listed on Appendix II. Initially, the American black bear was not listed under CITES.

It is virtually impossible to distinguish between the gall bladders of different bear species without conducting a laboratory test. For this reason, customs officials have difficulty in differentiating between gall bladders from Appendix I listed bears and other species. This 'look-alike' problem is of great concern because it provides an opportunity to trade in gall bladders of Appendix I bears on the pretence that they were obtained from unprotected species. As long as the American black bear remained unlisted, traders could 'launder' gall bladders from illegally poached Appendix I species through the US and Canada.

In recognition of the look-alike problem Canada listed its population of American black bears on CITES Appendix III. However this measure proved to be of little use as many Asian range states do not make provision for Appendix III species in their domestic legislation. The continued problem of laundering through North America prompted both the US and Canada to list the American black bear on Appendix II at the eighth COP in 1992.

What effect did this have? By September 1995 only three CITES permits had been issued to export American black bear gall bladders from the US, and ten permits had been issued in Canada. This was certainly not representative of the level of trade in bear products taking place through those countries. Most trade in bear parts takes place informally between individuals who are reluctant to comply with customs formalities. The only really noticeable effect of the Appendix II listing was to increase the administrative burden on all those involved in Canada's trophy-hunting business: hunters, outfitters, managers and customs officials. The Appendix II listing of the American black bear has had no discernible effects on bear conservation generally.

At the tenth COP Jordan, Bulgaria and Finland submitted a proposal to list European and Asian brown bear populations on CITES Appendix 1, but this was rejected. However, a resolution was passed that calls for Parties to reduce illegal trade through the usual means: strengthening laws, increasing penalties and improving law enforcement training. The resolution also suggests that Parties seek ways to reduce consumer demand for bear products.

Lessons

The case of the bear trade highlights several problems with CITES. First, it suggests that Appendix III listings may have little value, because they are ignored by some consumer nations. Second, it indicates some of the difficulties with an Appendix I listing for all bear species. If this were done bear farms in China would remain operational, and legal bear hunting would continue in various range states. However, legal hunters would be discouraged from harvesting gall bladders (other than to sell to a few selected domestic markets). By further restricting the supply of bear products to the market, their price would most certainly increase. This would encourage further bear farming within China and would create additional incentives for poaching in all range states. Moreover, listing all bears on Appendix I would still not solve the look-alike problem, because bear gall bladders are virtually indistinguishable from the gall bladders of other animals such as pigs. Indeed, many 'bear' gall bladders available on the market are in fact fakes from pigs, cows and other animals. It might appear that the only solution to this problem is to list all look-alike species on Appendix I. But as this would involve listing species such as the domestic pig this is obviously impractical.

The well-established practice of bear farming in China presents a further challenge to CITES. The Chinese government has indicated that it believes that bear farms are the only practical solution to the bear conservation problem, and is clearly reluctant to close them down. In any event, closing all bear farms would probably have disastrous consequences for wild bears, because the resultant supply shortage would almost certainly trigger a drastic price increase, which would fuel another surge in poaching activity. Most farmed bears are Appendix I listed species, so China is not allowed to export these, unless it registers its farms as captive breeding facilities under CITES. Thus far, the Chinese government has not applied to register its farms, because of fairly vigorous opposition to this idea. However, there is clear evidence that bear farming has substantially reduced the domestic market price of wild-harvested bear gall bladders in China, and there can be little doubt that if China were allowed to export farmed bile, prices would fall elsewhere too, with beneficial effects for wild bears everywhere. Unfortunately, bear farming raises significant animal welfare considerations and there is a degree of conflict between what is best for the welfare of individual farmed bears, and what is best for the conservation of wild bears. CITES is not designed to deal effectively with this conflict.

One possible way forward is suggested by the look-alike problem. This provides an opportunity to create a credible legal supply source of officially authenticated bear gall bladders that could effectively out-compete many illegal supply sources. Hong Kong has already introduced a system of authentication, with positive results. If a consumer authentication system could be linked to an appropriate mechanism to retrieve gall bladders from legally hunted bears, this could have positive benefits all around – for hunters, conservationists, consumers and animal welfarists eager for an alternative to bear farming. But it is not clear that the convention is appropriately structured to facilitate the development of such a system.

CONCLUSIONS

The four case studies discussed above have certain similarities. All four involve large charismatic mammal species that yield high-value products that are in great demand in Asian markets. Three of the examples involve products that are sought after as essential ingredients in traditional Asian medicines, and for which substitutes are not readily accepted (this is especially true of rhino horn). All these products are relatively easy to smuggle; they can be reduced to fairly small sizes and easily concealed; some (eg bear gall bladders) also resemble other products for which trade is legal, thereby complicating the task of customs officials.

How well did listings on CITES Appendices I, II and III perform for these case studies? The only species that seems to have benefited from an Appendix I listing is the African elephant. However, as discussed above, this listing would probably not be economically sustainable in the long term, as elephant numbers continue to increase to problematic levels. Appendix I listings have not stopped illegal commercial exploitation of rhinos, tigers and bears.

The Appendix II downlisting appears to have worked for the southern white rhino, but we can probably attribute this to good domestic management and field protection rather than to CITES. An Appendix II listing did not appear to work for the African elephant and the listing of the American black bear for look-alike reasons has been largely ignored by traders of bear products, while creating unnecessary additional complications for legal trophy hunters. Similarly, an Appendix III listing of the American black bear seemed to have little positive effect.

The system of listing species on different appendices is problematic. If a particular species is split-listed it creates an opportunity to launder products through the jurisdiction with the most lenient regulations. The only real solution is to list the species on one appendix over its entire range, as has been done for the African elephant. The problem with this is that, inevitably, the entire species must be accorded the strictest status (ie, Appendix I) thereby penalizing range states with good management systems who are both capable and keen to engage in legal trade. CITES tends to benefit those range states with poor management systems and inadequate field enforcement, thereby creating perverse and inappropriate incentives.

The Appendix II system assumes that wildlife trade is a formal sector activity, and that all traders have an incentive to trade through legal and formal channels if their product was legally obtained. This is not so; most wildlife trade takes place through the informal sector, through traders who are keen to avoid customs duty and other taxes, and thus have incentives to under declare their product shipments, if they declare them at all. Customs officials are not conservationists, but CITES places the full burden on them to catch offenders, without providing them with any real incentive to do so. One of the greatest challenges facing CITES is to create incentives for commercial exploitation and trade to take place through legal channels, regulated and monitored by people with a vested interest in conservation.

The recent, carefully restricted Appendix II listing of selected African elephant populations is a tentative step in this direction. However, it involves a complex set of procedures, checks and balances. There are insufficient resources within the CITES system to develop similarly intricate measures for all other species.

The most serious shortcoming of CITES is its narrow focus on restricting trade. Trade itself is not bad for conservation. There are many examples of species whose conservation status has been greatly enhanced by commercial exploitation, such as the southern white rhino. The future of successful conservation lies in recognizing instances where trade can be beneficial to a species, and creating a mechanism that encourages sustainable use and legal trade, while discouraging unsustainable and illegal exploitation.

REFERENCES

Barbier, E, Burgess, J, Swanson, T and Pearce, D (1990) *Elephants, Economics and Ivory*, Earthscan, London

Milliken, T, Novell, K and Thomsen, J (1993) *The Decline of the Black Rhino in Zimbabwe*, TRAFFIC International, Cambridge

't Sas-Rolfes (1995) *Rhinos: Conservation, Economics and Trade-offs*, IEA Environment Unit, London

Chapter 8

Conservation of the Nile Crocodile: Has CITES Helped or Hindered?

Henriette Kievit

INTRODUCTION

The main goal of CITES is to protect species of wild fauna and flora against over-exploitation through international trade. To achieve this, the convention regulates international trade in the products of those species for which trade poses an extinction risk. Since 1973, thousands of species have been listed on the appendices of the treaty. For most species it is not clear that listing has been beneficial. It is difficult to find examples of endangered species that have recovered to such an extent that they can either be moved to an appendix that allows for more trade, or be removed from the appendices altogether. This has led some to conclude that CITES has not worked.

Others defend CITES vigorously and point to crocodiles as an example. According to this school of thought, crocodilians stand out as the greatest conservation success story of the last quarter of a century. As a consequence of large scale, uncontrolled hunting from the 1940s to the beginning of the 1960s, crocodile populations became severely depleted in many parts of the world. By 1969, all 23 species of living crocodilians were either already endangered or drastically decreasing in numbers. But in 1994, after 25 years of conservation efforts by CITES member states, eight species were sufficiently abundant to sustain a well-regulated commercial harvest; eight species were safe from extinction but could not sustain a harvest; seven remained critically endangered.

The critics of CITES respond that while crocodiles provide an impressive demonstration of the effectiveness of conservation through sustainable utilization, they do not illustrate the efficacy of CITES. It is claimed that the success of crocodile conservation is actually due to a departure from the typical CITES remedy. This view is endorsed in this chapter. CITES, as originally conceived, took its lead from Western, protectionist approaches to wildlife and did not allow a role for sustainable use in conservation. But the key to the success of crocodile conservation has lain in the promotion of sustainable-use projects. This is illustrated by the case of the Nile crocodile (*Crocodylus niloticus*), which will be the major focus of this chapter.

THE EARLY DAYS OF CITES AND CROCODILES

CITES was established in 1973 at a plenipotentiary conference in Washington. At that time, although there was widespread acceptance that action should be taken to save species, there was little if any information available about the status of many of the world's wild species and only a rudimentary understanding of the nature of the threats and the sorts of action which might help. The assumption in CITES is that international trade is the main threat to the survival of species. Threatened species are therefore to be listed on one of the three appendices to CITES. Appendix I is for the most threatened species and imposes an almost complete ban on international trade in any species so listed. Appendix II is for species that are not necessarily now threatened with extinction but which may become so unless trade is regulated. An Appendix II listing imposes some regulation of trade.

At the time of the Washington conference CITES had not developed any explicit criteria to guide listing decisions. Nevertheless, at that meeting 1,100 species were placed on the appendices. They were listed according to the advice of Western experts. The 27 developing countries that attended the plenipotentiary meeting had little voice at the technical level (Hutton and Games, 1992). It was only at the first Conference of the Parties, held in Berne, Switzerland in 1976, that, with the passing of Resolution Conf 1.1 and Resolution Conf 1.2, criteria were established for the listing of species on the appendices. The criteria contained in those resolutions became known as the 'Berne criteria'.

The Nile crocodile was one of those species listed on Appendix I in 1973 without reference to any criteria. The justification for this listing has since been the subject of much debate. Producer countries of crocodile products have invested enormous effort to achieve a transfer of the crocodile populations to Appendix II. This would enable them to treat crocodiles as an economic asset.

On the face of it, the Appendix I listing meant that no trade in crocodile parts was possible. However, in practice, trade in crocodiles was able to continue through several different mechanisms. First, CITES had no jurisdiction over non-Parties and trade could continue among non-member states. During the 1970s important producer and consumer countries, such as Zimbabwe (then Rhodesia), France and Italy were not members. Second, the convention allows countries to make reservations in respect of the listing of particular species on Appendix I or II. A country which makes a reservation is not bound by that listing decision. These reservations became important as more countries joined CITES. France and Italy, for example, entered reservations when they acceded to the convention in the late 1970s. The fact that Italy imported 58,871 skins during the 1979–82 period showed the importance of these reservations (Caldwell, 1984). Japan was another consumer country that held a reservation on the Nile crocodile. Producer countries that entered reservations on acceding to CITES were Botswana (1978), Zambia (1981), Zimbabwe (1983) and Sudan (1983).

The third mechanism that appeared to enable trade in Nile crocodile products was Article VII, paragraph 4 of the convention. Article VII(4) allowed specimens of Appendix I species which were bred in captivity for commercial purposes to be treated as if they were Appendix II specimens. This form of exploitation was not considered to threaten the survival of species because it was supposed to have no impact on wild populations; it was even believed that it could relieve the pressure on wild populations.

However, each of these loopholes was gradually tightened. First, as more countries acceded to the convention, scope for trade among non-members declined. Second, member countries came under pressure to withdraw their reservations. Finally, the opportunity for trade in crocodiles seemingly offered by Article VII(4) was closed. Before the second COP in 1979 there was some ambiguity about the scope of Article VII(4). At issue was the meaning of 'bred in captivity'. Some countries interpreted this broadly to include the rearing in captivity of specimens taken from the wild when young. The Cayman turtle farm had been trading in turtles on the assumption that this form of captive rearing was an acceptable activity under Article VII(4). Other countries disputed this, maintaining that captive breeding ruled out taking specimens from the wild. The situation was clarified when the second COP, held in 1979 in San Jose, Costa Rica, adopted Resolution Conf 2.12. This defined 'captive breeding' in such a way as to effectively rule out trade in specimens taken from the wild and then reared in captivity. The term 'ranching' was to be assigned to activities that involved captive rearing.

With the gradual closure of these loopholes attention turned to ways in which crocodiles could be transferred from Appendix I to Appendix II. At this point the concerns of those Parties who wished to trade in crocodiles coincided with more general worries about the Berne criteria felt by some Parties. As so many species had been listed prior to the adoption of the Berne criteria, and it was felt that mistakes had been made in this process, the mechanism for deleting species or transferring them from Appendix I to Appendix II was of crucial importance. Deletion and transfer was dealt with in Resolution Conf 1.2, but this resolution made it very difficult to do this. It requires evidence that the species has recovered sufficiently to allow trade and that it can withstand the exploitation resulting from the removal of protection. However, as there had often been no clear data on the status of species at the time of listing, it was often impossible to prove that the species had recovered – a catch-22 dilemma. It was similarly difficult to show, in advance, that the species could withstand exploitation. The only crocodile population ever downlisted pursuant to Resolution Conf 1.2 was the American alligator (*Alligator mississippiensis*) in 1979.

The difficulties of satisfying the requirements of Resolution Conf 1.2 were already realized at the time of the second COP in 1979. A general response to these difficulties came in the form of a proposal from the US which resulted in the passing of Resolution Conf 2.23. This addressed the issue by establishing special criteria for the deletion of species without the application of the Berne criteria. However, the final wording of the resolution was unclear and made the resolution virtually useless (Wijnstekers, 1992, p 225).

Another response dealt specifically with species such as crocodiles that could be ranched. It was recognized that some ranched populations of Appendix I species, such as the sea turtle (*Chelonia mydas*) and Nile crocodile, could withstand a certain level of exploitation and might, in fact, benefit from controlled utilization. Several countries informally expressed concern about the extent of the restrictions on the trade in ranched animals. They argued that unless the disposal of surplus stocks of ranched Appendix I species could be made commercially viable, conservation authorities in some countries would be unable to prevent the land reserved for such species from being used for other income-generating activities. This would be contradictory to the intentions of CITES and it was said that it would lead to the weakening of the convention. In the light of this discussion the second COP established a committee, known as the Ranching Committee, to examine these issues and to report before the third COP.

THE RANCHING COMMITTEE AND THE RANCHING RESOLUTION

The Ranching Committee needed to establish criteria under which ranched specimens of Appendix I species could be transferred to Appendix II. It failed to reach agreement before the third COP in New Delhi in 1981 and at the conference a working group was formed to address the issue. Although at that time Zimbabwe had only observer status at CITES, representatives of that country contributed to the discussion, emphasizing the conservation benefits of ranching operations. The working group agreed to the wording of a resolution that was subsequently adopted by consensus by the conference. Resolution Conf 3.15 allows for the transfer of individual populations of Appendix I species to Appendix II for the purposes of ranching, where those populations are no longer endangered and a number of other conditions are satisfied.

What effect did this resolution have on crocodile conservation? In the case of Zimbabwe, the effect was positive. When Zimbabwe joined CITES in 1983 it had a reservation on the Nile crocodile that enabled the country to continue its trade in crocodile skins with Italy, France and Japan. Zimbabwe wanted to continue its successful sustainable-use conservation project for the species. Zimbabwe's project was based on ranching; eggs were collected from the wild and animals were reared in captivity, after which a small fraction was returned to the wild and the products of the remainder were used for international trade. The rearing stations placed an economic value on crocodiles in the wild, thereby contributing markedly to their protection. These stations did much to educate the public towards tolerating the species in areas where there was prior tension. As a result, at the beginning of the 1980s, the species was in no danger in Zimbabwe.

The CITES Secretariat, communicating the dismay of other states with Zimbabwe's reservation, suggested that Zimbabwe should attempt to get its Nile crocodile population reclassified on Appendix II pursuant to Resolution Conf 3.15. So, before the fourth COP in Gaborone, Botswana in 1983, Zimbabwe approached other Parties and interest groups on two occasions. First, at the 1982 Symposium on Crocodile Conservation and Utilization at Victoria Falls, CITES member-states and NGOs were informed of Zimbabwe's conservation strategy. Convinced of the merits of Zimbabwe's conservation programme, symposium members adopted a resolution supporting the application of Zimbabwe to transfer its Nile crocodile population from Appendix I to Appendix II. Second, during the 1982 meeting in Nairobi of the Committee of the African region for the Ten Year Review of the Appendices, the proposal was also discussed. Although the proposal to transfer the entire species from Appendix I to Appendix II was strongly opposed, Zimbabwe's ranching proposal was unanimously supported. Lastly, IUCN also approved of Zimbabwe's proposal. As a result, Zimbabwe's move to have its crocodiles transferred to Appendix II in accordance with Resolution Conf 3.15 was endorsed at the Gaborone Conference.

One important consequence of the downlisting of the Zimbabwean Nile crocodile was the general acceptance of the conservation benefits of ranching for this species. Another was that it set an example of 'split-listing'. This countered the tendency of CITES to list all the populations of a species on Appendix I, when the population in one country was considered to be threatened. It also illustrated how a few countries could influence the direction of CITES and how reservations could be a valuable mechanism for achieving progress in CITES. Zimbabwe was able to demonstrate the viability of its crocodile ranching because it continued to trade in crocodile products under the reservation it held. But while Zimbabwe's case was seen as an example for others to follow, in practice, Resolution Conf 3.15 did not make this easy. The criteria for ranching, although not requiring proof that the population had recovered, demanded such strict controls on the management of wild populations and the conduct of the ranching operation that many countries without a long history of crocodile management would have great difficulty in fulfilling them. At the time, few such well-managed programmes existed. In Florida and Louisiana, State-managed harvest programmes were developed for the American alligator. Venezuela experimented with sustainable-use programmes for the Orinoco crocodile (*Crocodylus intermedius*) and the spectacled caiman (*Caiman crocodylus*), as did Papua New Guinea for the New Guinea crocodile (*Crocodylus novaeguineae*) and the saltwater crocodile (*Crocodylus porosus*). But in Africa, Zimbabwe was the only country with a well-developed ranching programme for the Nile crocodile.

For other countries with an interest in ranching, but with no existing ranching projects, the resolution did not prove to be of much use. Beginning ranching programmes without being able to trade the products until the viability of the operation was proven, imposed unbearable start-up costs on range states and was therefore not an option. Several African countries with abundant Nile crocodile populations found themselves in this position. Even though the benefits of ranching for conservation were recognized, the restrictions of Resolution Conf 3.15 deterred them from starting actual ranching programmes. At the Gaborone conference Botswana's proposal to downlist its crocodile population under Resolution Conf 3.15 was rejected because it lacked sufficient information on its conservation programme and on the status of the species. For these same reasons, Australia and Surinam withdrew their proposals to downlist the saltwater crocodile (*Crocodylus porosus*) and the sea turtle (*Chelonia mydas*) respectively, before it came to a vote. Proposals from Mozambique and Madagascar to downlist their individual Nile crocodile populations under the Berne criteria were also rejected. Yet, a study on the Nile crocodile carried out on the behalf of the CITES Secretariat, demonstrated that the species was sufficiently abundant in African countries such as Malawi, Zambia, Tanzania and Botswana to allow sustainable use. So, even after the passing of Resolution Conf 3.15 and the downlisting of Zimbabwe's crocodiles in accordance with that resolution, there was still pressure to make it easier to downlist other populations in order to allow ranching to take place.

SPECIAL CRITERIA FOR THE TRANSFER OF SPECIES FROM APPENDIX I TO APPENDIX II

In 1984 these problems were discussed in Brussels at the first meeting of the CITES Technical Committee. African range states that did not want to, or could not downlist their species under Resolution Conf 3.15 felt they had to fight to get species downlisted under the Berne criteria. They argued that the catch-22 of showing changed status often posed an insurmountable obstacle which frustrated their attempts. Moreover, Italy and France had withdrawn their reservations on the Nile crocodile in order to align themselves with the other European Community countries. This resulted in the loss of two important markets for the African range states. The desire to resume trade with these countries fueled their campaign to get the Nile crocodile onto Appendix II.

The Brussels meeting recognized the need for an alternative procedure to allow utilization of wild populations while information was gathered to satisfy the criteria of Resolution Conf 1.2 or Resolution Conf 3.15. After a lengthy discussion it was concluded that the basic principle of the Berne criteria, the requirement that downlisting should not threaten the species, had to be maintained. But, it was also agreed that quotas could be used to ensure that the exploitation of a species would not threaten its survival.

At the fifth COP, held in Buenos Aires, Argentina in 1985, the draft resolution of the Technical Committee was amended by a small working group consisting of Canada, Malawi, Sudan, Switzerland, Zimbabwe and the US. The US eventually proposed Resolution Conf 5.21, which was entitled: 'Special criteria for the transfer of species from Appendix I to Appendix II'. This was adopted without objection. Resolution Conf 5.21 acknowledged that the Berne criteria for the transfer of species from Appendix I to Appendix II were very difficult to fulfil. It also acknowledged that certain species had been listed prior to the Berne criteria and had probably never met the criteria, or had recovered since their inclusion in Appendix I. It therefore established a quota system under which Parties could export products of species which could withstand a certain level of commercial exploitation and could not be downlisted to Appendix II under the Berne criteria or Resolution Conf 3.15. It gave countries four years to gather the necessary data needed to get their populations transferred to Appendix II pursuant to the Berne criteria or the Ranching Resolution.

Under Resolution Conf 5.21, the populations of the Nile crocodile in Cameroon, Congo, Kenya, Madagascar, Malawi, Mozambique, Sudan, Tanzania and Zambia were transferred to Appendix II in 1985, followed in 1987 by the Nile crocodile population in Botswana. To these were added the populations of the Nile crocodile in Ethiopia and Somalia (1989) and those of South Africa and Uganda (1992). Populations in Cameroon, Congo and Sudan were assigned zero quotas in 1989 and were, therefore, effectively prohibited again from trade. They were finally retransferred to Appendix I in 1992. In 1989, the Nile crocodile populations of Botswana, Malawi, Mozambique and Zambia were retained in Appendix II under the terms of Resolution Conf 3.15, and were therefore no longer restricted by export quotas. The populations of Ethiopia, Kenya and Tanzania followed in 1992.

Two factors played an important role in both the adoption of the Resolution Conf 5.21 in 1985 and in the subsequent adoption of quotas for the African range states. First of all, African countries raised their voices and criticized the attitude of Western countries. On behalf of the African group, Malawi submitted a statement denouncing the tendency of Western countries to regard the findings of the scientific authorities of African Parties to CITES with suspicion. The statement also urged other Parties to accept the proposals of African countries to downlist individual Nile crocodile populations. Secondly, African countries united and voted as one, which considerably strengthened their position.

Developments After the Buenos Aires Conference

In the period following the Buenos Aires Conference the quota system was generally seen as successful, though not perfect. IUCN endorsed the crocodile quota system as being within the world conservation strategy concept of sustainable use of resources. The conservation status of the Nile crocodile was seen as safe.

At the seventh COP, held in Lausanne, Switzerland in 1989, Resolution Conf 5.21 was replaced by Resolution Conf 7.14. Resolution Conf 7.14 also provides for a temporary mechanism to transfer species to Appendix II that were incorrectly listed in Appendix I. It acknowledges that Resolution Conf 5.21 had proved to be useful and that its principles should be kept as an interim mechanism for the transfer of taxa from Appendix I to Appendix II. Resolution Conf 7.14, however, defines the requirements that should be fulfilled in order to downlist species more effectively. New requirements were that species should be non-migratory, the Party should have a scientifically based and well-documented programme, the products should be adequately marked and the Party should have no reservation on the species. Since the passing of Resolution Conf 7.14 there have not been any major changes to the CITES policies on crocodiles, although there have been on-going attempts to refine the tagging systems for identifying crocodile skins.

Conclusion

This chapter has shown how CITES, after originally listing the Nile crocodile on Appendix I, did eventually come around to supporting a policy of sustainable use. Crocodile ranching has proved to be a remarkably successful way of promoting crocodile conservation. Populations of the Nile crocodile have been recovering since the 1980s in Botswana, Ethiopia, Kenya, Malawi, Mozambique, Tanzania, Zambia and Zimbabwe. However, it took a long time for CITES to change its position. What was needed, after the Appendix I listing in 1973, was a viable mechanism for transferring crocodiles to Appendix II in order to allow ranching to take place. The original resolution covering such transfers was Resolution Conf 1.2, but this proved almost impossible to satisfy. Resolution Conf 3.15 attempted to solve the problem but it turned out to be virtually useless for African range states which had not yet established ranching programmes. It was only with the passing of Resolution Conf 5.21 that CITES arrived at a position favourable to crocodile conservation and started to function as a force to improve the management of these species.

The key to the success of ranching programmes is that it puts an economic value on wild crocodile populations (which are the source of eggs) and thus makes it a competitive land-use option. People who live with the animals can share in the benefits accruing from ranching programmes. But as long as crocodiles remained on Appendix I then crocodiles in the wild lacked economic value. There were no incentives to counter demands to eradicate crocodiles and use the crocodile habitat for other purposes. In the developing world, wildlife is competing with humankind for limited resources. Denying wildlife a commercial value denies it the opportunity to compete successfully with alternative land use practices.

In part, the dispute within CITES about crocodiles was a debate between the differing views of conservation held in the developed and developing world. Developed countries usually make decisions on trade in wildlife products from positions of affluence and are influenced by protectionist and animal rights groups. Developing countries usually make decisions from a position of poverty and are influenced by the immediate needs of their inhabitants. This predisposes them towards conservation solutions that allow people to benefit from the use of wildlife. The fact that developing countries were able to win the argument in the case of crocodiles may have been due to the non-charismatic nature of the species. Crocodiles have not attracted the sort of attention that Western environmentalists have given to other species such as elephants and this has made it easier to establish a policy based on sustainable use.

REFERENCES

Caldwell, J (1984) *All CITES Trade in Crocodylus Niloticus*, CITES Secretariat, Lausanne

Hutton, J and Games, I (1992) *The CITES Nile Crocodile Project*, CITES Secretariat, Lausanne

Wijnstekers, W (1992) *The Evolution of CITES*, CITES Secretariat, Lausanne

Chapter 9

Are All Species Equal?
A Comparative Assessment

Grahame J W Webb

INTRODUCTION

CITES was drafted in the early 1970s. At that time, wildlife conservation was becoming a highly public and politically significant issue in many countries. Advocates of conservation increasingly saw their goal as the protection of wildlife from any form of consumptive use. While international trade lay at the root of a number of wildlife conservation problems, the aim of some people and organizations associated with CITES was to stop, rather than to control, that trade. Much has changed since then. A number of the Parties to CITES have recognized that international trade, if properly controlled, can create powerful incentives for conservation. The concept of using species sustainably, although long applied to game species around the world, has become more widely recognized as a potential conservation tool. These insights represent changes in the basic philosophy of wildlife conservation within CITES. While some have embraced those changes, however, others have opposed them bitterly.

This underlying philosophical conflict in CITES is nowhere more apparent than in relation to charismatic species. Charismatic species are those which wild species that, for various reasons, are regarded by the public as special in some way. Such a view may arise because of their size (elephants and whales), their assumed intelligence (dolphins), their harmless non-predatory nature (sea turtles), their big brown eyes (seals), or simply their warm cuddly appearance (pandas). At times the reasons why a species is considered charismatic are simply unknown: the public simply identifies them as such.

Charismatic species create significant problems of consistency within CITES. The text of the convention did not establish different rules for charismatic and non-charismatic species. Crocodiles and sea turtles should be treated similarly, for example, despite the charismatic nature of the other. The convention assumes that similar wildlife populations function in similar ways and can be affected by trade and safeguarded in similar ways. The text looks towards science and technology for its answers, rather than to emotion.

In reality, the work of the Parties at successive COPs is heavily biased towards charismatic species and they are not treated in the same way as non-charismatic species when it comes to the final voting. Proposals involving crocodiles are dealt with far more swiftly and expediently than those involving sea turtles. The very text of the convention has been interpreted differently for turtles and crocodilians. For example, the guidelines for ranching sea turtles arguably require unobtainable evidence, whereas the information required for ranching crocodilians is straightforward. The precautionary principle is adhered to far more vigorously for sea turtles than it is for crocodiles. Even the independent advice the Parties receive on proposals from organizations such as IUCN, varies greatly between turtles and crocodiles.

This chapter examines the way in which sea turtles and crocodilians have been treated within CITES. It argues that, despite differences in the extent of their movements, the two groups of species are very similar biologically and they have similar conservation and management needs. As such, they should be treated in a much more similar way by the Parties. To date, this has not occurred.

CROCODILES AND TURTLES

Attempts to promote crocodiles as charismatic species have not been successful for obvious reasons. They eat people, prey on domestic stock, compete with people for food and have long been regarded as evil vermin. Sea turtles, on the other hand, are usually seen as harmless and highly charismatic. Yet, to those who live with these animals, both are sources of food and products for trade. Indeed, the consumptive use of both may well have been occurring for hundreds of thousands of years.

In the 1970s, pressure mounted to use CITES as a tool for stopping all international trade in sea turtles and crocodilians. This pressure was consistently opposed by some Parties in relation to crocodilians. In the early 1980s, the conservation benefits of trade in wild crocodilian populations was recognized by most Parties and crocodilian species started to be transferred from Appendix I to Appendix II for harvesting from the wild. At each COP since then, the Parties have approved more proposals to harvest wild crocodiles and have registered more commercial captive breeding operations for crocodiles on Appendix I. The IUCN–SSC Crocodile Specialist Group, which provides Parties with advice on proposals concerning crocodilians, generally works with countries proposing particular uses.

In contrast, the Parties have not approved any transfer of sea turtles from Appendix I to Appendix II, and they have not agreed to register a single captive breeding centre. Indeed, the prolonged attempt by the Cayman Islands to have their green turtle (*Chelonia mydas*) farm-registered for ranching and captive breeding is a remarkable saga of determined political opposition, often based on spurious, or at least questionable, technical grounds. With sea turtles, the IUCN–SSC Marine Turtle Specialist Group has been reluctant to assist any Parties proposing consumptive use.

This dichotomy in the way in which the Parties to CITES have treated the two groups of long-lived reptiles – the non-charismatic crocodiles on the one hand and the charismatic sea turtles on the other – remains steadfastly in place today. At the tenth COP, held in Harare in 1997, there were five proposals for increased use and trade in crocodilians, all of which went to the Parties with varying levels of conditional support from the Crocodile Specialist Group. There was also one proposal for trade in turtles, which had no support from the Marine Turtle Specialist Group. The fate of the crocodilian proposals was as follows:

- Honduras proposed the registration of a captive breeding farm for American crocodiles (*Crocodylus acutus*), the founder stock of which had been taken from the wild. The Parties approved it.
- Argentina proposed the transfer of its wild population of broad-snouted Caimans (*Caiman latirostris*) from Appendix I to Appendix II for ranching. The Parties approved it.
- Madagascar proposed the transfer of its wild population of Nile crocodiles (*Crocodylus niloticus*) from Appendix I to Appendix II for ranching. The Parties approved it.
- Uganda proposed the transfer of its wild population of Nile crocodiles (*Crocodylus niloticus*) from Appendix I to Appendix II for ranching. The Parties approved it.
- Tanzania proposed the transfer of its wild population of Nile crocodiles (*Crocodylus niloticus*) from Appendix I to Appendix II on the basis of a quota allowing 1,100 animals to be killed and exported. The Parties approved it.

It was a different story with sea turtles. Cuba proposed that the population of hawksbill turtles (*Eretmochelys imbricata*) in Cuban waters be transferred from Appendix I to Appendix II, so that Cuba could trade in the shell derived as a byproduct of its domestic harvest (which itself had been reduced by 90 per cent), and continue to develop ranching and trade in the legal stockpile accumulated from its domestic harvest. The proposal was extensive and well-supported with research data. Despite winning the majority of votes, the proposal failed to reach the two-thirds majority required for a transfer from Appendix I to Appendix II. Once again, the Parties had sanctioned the use of crocodilians but blocked the use of sea turtles.

If these two groups of animals were vastly different in their biology and population dynamics, then completely different approaches to their conservation and use might be warranted on the basis of scientific and technical considerations. But in fact, they share many characteristics.

HOW BIOLOGICALLY SIMILAR ARE CROCODILES AND SEA TURTLES?

At a very basic level, sea turtles and crocodiles are vertebrate reptiles with a long evolutionary history. They are both large, semi-aquatic and capable of living in salt water, three traits that are exceptional among living reptiles. Both are heavily armoured to protect their internal organs. Their ancestors were clearly robust and able to survive the extinction of most large reptiles 65 million years ago. Both groups are ectothermic (cannot control their body temperature through endogenous heat production) and so ambient temperatures have a profound effect on their growth and metabolism.

Obligate oviparity characterizes both groups; in neither sea turtles nor crocodiles are there indications of embryos having prolonged development within the oviducts of their mothers. Both groups nest in restricted nesting areas. They typically bury their eggs in terrestrial locations, out of the water. They lay large numbers of eggs, which take long periods of time (at least two months) to develop. The parents do not move the eggs during incubation. The eggs are vulnerable to mortality from inundation, predation and other causes. Embryo development rates vary with changes in the ambient temperature, moisture and gaseous environment. Both groups have their sex determined by incubation temperature and in neither group is sex reversible.

During their lifetimes, female crocodiles and sea turtles produce many times the number of potential offspring needed to replace themselves. Mortality rates between egg laying and the attainment of maturity are high (more than 99 per cent in many species). The sex ratio of wild populations is often not 50 per cent males and 50 per cent females. Rates of growth and ages at maturity vary between individuals and areas. Maturity often takes 10–20+ years. The relationship between size and age usually includes a prolonged period of essentially linear growth. Both groups have long potential life spans. Sea turtles and crocodiles have refined navigation skills, allowing them to return to specific areas from considerable distances. Both groups are known to make significant movements from their natal site. Marine crocodiles, like sea turtles, can make long-distance voyages around coastlines, through island chains, and out to sea. Nevertheless, this is at a much lesser level than characterizes the average sea turtle, which regularly makes long distance voyages at sea between nesting and feeding areas.

So, despite one group having little charisma and the other a lot, and one group being much more mobile at sea than the other, crocodilians and sea turtles are very similar. They share a great many biological traits and have similar population dynamics. The increased mobility itself suggests that different administrative difficulties may need to be overcome, rather than that there are major biological differences.

How Similar Are the Conservation and Management Needs of Sea Turtles and Crocodilians?

The eggs and meat of sea turtles and crocodiles have been used for food by people for tens of thousands of years. In both cases, the wild species convert food sources that are unavailable to people into protein sources that are available. The use of sea turtles and crocodiles has strong traditional and cultural associations for many people, particularly impoverished people who live in coastal areas and freshwater wetlands. Products derived from crocodiles and sea turtles have long been traded at local, national and international levels, and, therefore, have had economic significance to those who live with the resource. Through international trade, some products made from crocodiles and sea turtles now have economic and cultural value to people outside producing areas.

The densities of the wild populations of both groups can be significantly reduced through overharvesting, particularly for trade. However, both groups are tenacious survivors: no known species of crocodile or sea turtle has become globally extinct in recent history, despite a few local extinctions. Yet, most species in many areas have been intensely utilized in the not too distant past. Even in heavily harvested situations, economic extinction usually occurs well before biological extinction.

The nesting areas of both crocodiles and sea turtles are vulnerable to economic development: agriculture in the case of crocodiles; coastal development in the case of sea turtles. Bycatch in commercial fishing operations is a significant problem with both groups. The impact of intense egg harvesting may take a long time to be reflected in adult populations and it is biologically less significant than the impact of harvesting adults. In both groups the removal of one adult female is equivalent to the loss of many eggs.

Density-dependent changes in survival rates (ie, increased survival with reduced densities) are strongly suspected to compensate for the harvest of crocodiles, and the same may be true of sea turtles. In both groups definitive data on survival rates are hard to obtain. However, less than 1 per cent of eggs in both groups appears to survive to maturity. Both species are generally amenable to ranching given a reasonable level of technological input. Eggs are typically easy to find, move and incubate using standard technology. Hatchlings can be raised successfully, but require strict attention to the raising environment. Risk of mortality decreases sharply after the first few months and growth rates can be enhanced with attention to temperature, diet and density. In general, sea turtle hatchlings appear less subject to stress than most crocodiles, but are more sensitive to water quality. Otherwise, they share many characteristics. Both groups are amenable to production through captive breeding, although this may not be desirable from a conservation viewpoint.

The status of both groups today is highly variable between species and areas. Thus, although sea turtles and crocodiles have populations numbering millions of individuals, they may be abundant in some areas, but greatly reduced in others. The status of populations (increasing, decreasing or stable) also varies between areas and often reflects local management. If their habitats are intact, depleted populations of both groups seem to have the ability to recover quickly if the threatening processes are removed.

In sum, sea turtles and crocodiles are used by people in similar ways, and their status is detrimentally affected by uncontrolled use.

With both groups, their long-term survival may be more threatened by the loss of nesting habitats than by the loss of foraging habitats or by harvesting. They are amenable to management in similar ways, and one might expect that they would be treated much more uniformly under the articles of the convention. The increased mobility of sea turtles, in a world in which territorial waters and the high seas are controlled by different authorities, requires a higher degree of international cooperation than is the case with crocodilians. But, given that CITES is an international cooperative treaty, it is hard to accept that the problems associated with this cannot be overcome.

How Similar Are the Organizations that Assess Proposals Involving Sea Turtles and Crocodiles?

For crocodiles and sea turtles, the two key advisory bodies to the Parties to CITES are the Crocodile Specialist Group (CSG) and the Marine Turtle Specialist Group (MTSG) respectively. These two well-established specialist groups, under the umbrella of the IUCN–SSC, have made a substantial commitment to advancing the conservation of their respective groups. The CSG and the MTSG both provide advice directly to Parties and to a number of other groups who independently advise Parties. However, the CSG and the MTSG have evolved differently and their approach to conservation problems is often different.

In the 1960s and early 1970s, when the CSG and the MTSG began to operate, both preservationist and sustainable-use approaches were accommodated by the two groups. Archie Carr, the mentor of many current sea turtle specialists, actively promoted the use of green sea turtles (*Chelonia mydas*) as production animals to provide food for people. Robert Bustard, working in Australia, did likewise with both crocodiles and sea turtles, with the full support of the CSG, the MTSG and the Australian Government. But by the mid-1970s, the CSG and the MTSG both became strongly preservationist. Outside the CSG and the MTSG the preservation philosophy of the day was reaching dizzy heights. The accepted wisdom was that the commercial, consumptive use of wildlife had no place in modern conservation, and it became increasingly common to treat people involved in wildlife trade with contempt and disrespect. There was little consideration given to cultures and traditions dependent on wildlife use. In the CSG and MTSG of the day, it resulted in two key assumptions. First, that legal trade stimulates illegal trade. Second, that production through ranching and captive propagation is neither biologically nor economically feasible.

Things started to change within the CSG in the mid-1980s. Crocodiles, like sea turtles, were protected, and this protection was clearly a key factor in causing their recovery in many areas. But with the recovery came increased crocodile attacks on people and livestock. Such attacks seriously threatened public support for a purely protectionist approach, and the CSG responded. Within five years, under the Chairmanship of Harry Messel, most CSG members recognized that sustainable use had an important role to play once the immediate risk of extinction had been averted.

In the MTSG, no such changes took place. Perhaps circumstances have not demanded a reappraisal of underlying philosophies, or perhaps there are sound reasons for rejecting the flexible approach of the CSG. However, the end result is that the two groups view their target animals quite differently. Advice given by the MTSG rarely, if ever, sanctions consumptive use or international trade, whereas advice from the CSG usually sanctions consumptive use and international trade as either short- or long-term goals.

With crocodiles, legal trade proved to be a significant deterrent to illegal trade, which is now markedly reduced around the world. Ranching and farming both turned out to be biologically and economically feasible when based on sound science and good technology. With sea turtles, it is often still assumed that legal trade will encourage illegal trade and that ranching and farming cannot be biologically or economically feasible. But, in the absence of evidence, they are no longer convincing arguments.

So, despite the biological similarities between sea turtles and crocodilians, and the fact that their conservation and management involve many parallel problems, the two key advisory groups to CITES approach the issues of consumptive use and trade very differently. This appears to reflect a difference in the underlying philosophy of conservation, as much as it does genuine scientific concerns. The fact that the advice ultimately emanates from the same organization (IUCN) creates both confusion and conflict.

CONCLUSION

The Parties to CITES have consistently treated sea turtles and crocodiles very differently and this seems to reflect the political sensitivities associated with the two groups. Sea turtles are highly charismatic and their consumptive use for food or trade is politically unacceptable in some countries. The Parties have not made a single significant decision that enhanced consumptive use or trade in sea turtles. Crocodiles, on the other hand, are seen to have no charisma and thus proposals involving consumptive use and trade have been regularly accepted by the Parties to CITES. This dichotomy in the way in which the Parties treat sea turtles and crocodiles does not appear to be based on the conservation needs of the two groups, although it is often argued on that basis. Rather, it seems to reflect the charismatic nature of sea turtles, and the non-charismatic character of crocodiles.

Chapter 10

Zimbabwe and CITES: Influencing the International Regime

Phyllis Mofson

INTRODUCTION

Over the last several years, Zimbabwe's attitude toward, and behaviour regarding, CITES has changed dramatically. A member since 1983 and brought into the debate by its opposition to the listing of the African elephant as an endangered species, Zimbabwe has emerged as a leader at the international level in promoting a sustainable-use paradigm within the CITES regime. The portrait of Zimbabwe's involvement painted in this chapter shows that multilateral agreements can be influenced significantly by Parties which feel unjustly treated, no matter how small these Parties may be in economic terms. Zimbabwe's approach to CITES has evolved from a position of angry protest, with threats of withdrawal, to the assumption of a leadership role, actively working within the CITES system to alter the nature and strategies of the organization. These efforts to initiate changes in regime procedures and principles have the potential to influence the future of CITES in far more permanent ways than the more visible and politically charged battles over individual species.

The examples of Zimbabwe and its neighbours in the Southern African Convention on Wildlife Management (SACWM) suggest that, rather than requiring the surrender of some degree of national sovereignty and power, membership in CITES can, in some cases, enhance sovereignty and the power to advance national interests in the international arena. Through learning to become an inside player in the CITES system, Zimbabwe has not only influenced CITES and its issue domain, but has empowered itself as an international actor.

CITES, ZIMBABWE AND ELEPHANTS

The African elephant was first listed under Appendix II in 1978, and a quota system on ivory was established in 1985. From 1979 to 1988, Zimbabwe exported about 100 tons of ivory to the international market. During the time the elephant was listed in Appendix II the illegal ivory trade swamped the legal trade in quantity, due to rampant poaching and smuggling throughout large parts of Africa. There is no real dispute that the population of African elephants continent-wide fell by as much as 50 per cent during the 1980s. Although herds in some southern African nations suffered, the range states of East Africa accounted for the majority of the elephants lost.

The discussion about moving the African elephant from Appendix II to Appendix I began in earnest in the 12 months prior to the seventh COP, which was held in Lausanne, Switzerland in 1989. It was prompted by the apparent failure of the ivory-export-quota system. This mechanism had initially been heralded with cautious optimism by the Parties to CITES and NGOs alike. After just two years, however, TRAFFIC concluded that the system had 'succeeded in controlling the movement of only 20 to 40 per cent of the total amount of ivory produced annually in Africa' (Thomsen, 1989, p 1). Furthermore, the quota system allowed for government-regulated trade in confiscated ivory. TRAFFIC estimated that because of this loophole, up to 70 per cent of the so-called legal ivory trade came from illegally hunted elephants.

In the autumn of 1988, in response to the failure of the CITES ivory-quota system to control the poaching of elephants, the US unilaterally adopted the African Elephant Conservation Act, which allowed it to impose a moratorium on imports of ivory from any country which participated in the ivory trade, or which was not a member of CITES. The US action, combined with an NGO-driven campaign in the US and Europe to change consumer attitudes towards ivory, resulted in a significant drop in the worldwide demand for ivory. Europe followed the US with a unilateral ban on commercial imports of ivory.

THE LAUSANNE COP

The Appendix I listing for the African elephant was proposed at the Lausanne COP by Austria, Gambia, Hungary, Kenya, Somalia, Tanzania, and the US. The proposal, based on recommendations from the CITES African Elephant Working Group and studies commissioned by the specially-convened Ivory Trade Review Group, was adopted over the objections of eleven countries, including the southern African countries, Botswana, Malawi, Zambia, South Africa and Zimbabwe. Four countries, including Japan, abstained from the vote. The southern African objectors and the People's Republic of China registered reservations to the listing within the prescribed 90-day period (China subsequently withdrew its reservation). Zimbabwe officially entered a reservation to the listing even before the meeting ended.

The Appendix I listing virtually ended what remained of the international commercial ivory trade. However, at the time of the listing the population of elephants in Zimbabwe and Botswana was considered stable or rising, with up to 120,000 animals in the contiguous population of the two countries. In light of these numbers and in view of their well-regulated trade in ivory, the southern African countries protested that the new listing was patently unfair.

Even before the Lausanne COP, it was acknowledged by many Parties and experts that a total commercial ivory trade ban would have adverse effects on conservation programmes in certain countries, and it was generally acknowledged that Zimbabwe had a legitimate and intractable problem. However, without a proven means either to distinguish legal from illegal ivory or to control legal trade tightly, the international consensus maintained that a total ban was necessary. Nevertheless, convinced by the arguments of Zimbabwe and other southern African countries that not all of Africa's elephant herds were endangered, the Parties adopted a proposal from Somalia to establish a review process for countries that wanted to have their elephants downlisted back to Appendix II.

In an attempt to gain greater control over their ivory exports, which they assumed would resume after the next COP, Zimbabwe, Botswana, Zambia, and Malawi formed an organization called the Southern African Centre for Ivory Marketing (SACIM). This was designed to be the 'sole exporting agency of ivory from members' (Riccuiti, 1993, p30). In August 1991, Zimbabwe hosted a SACIM workshop on the future of CITES. Although designed primarily as a forum for airing grievances, several influential documents emerged from the workshop, including a set of recommendations for new criteria for listing species in CITES Appendices and a report entitled *The Case for a New Convention on International Trade in Wild Species of Flora and Fauna.* This report contained a polemical argument to the effect that CITES is full of major defects, does not work, and needs to be replaced. On acceding to CITES following its independence, Namibia took a reservation on the African elephant and joined SACIM. Zambia, in contrast, withdrew from the agreement following a change in its government.

THE KYOTO COP

At the eighth COP held in Kyoto, Japan in March 1992, the SACIM countries submitted a joint proposal to downlist their elephant herds to Appendix II; however only two, Zimbabwe and Botswana, met the Somali Amendment conditions set at Lausanne for downlisting their herds. Partly as a result of this, but also in an effort to allay the fears of other African Parties over the resumption of an ivory trade, the SACIM countries amended their proposal during the conference to include a zero-quota on ivory sales, a removal of their reservations, and an automatic return of their elephant populations to Appendix I at the next COP if they could not present acceptable plans for ivory marketing and elephant management by that time. Despite these concessions, however, opposition remained overwhelming, and the proposals were finally withdrawn. In his speech withdrawing the SACIM resolution, Botswana's Minister of the Environment said, 'We are extremely perplexed... It seems to us that the goalposts have been moved... We will review our participation in CITES as soon as we have reported to our respective governments'.

Most observers concede that Zimbabwe had a legitimate case for downlisting. Its elephant herds were not endangered and were being competently managed. So, its assumption that a downlisting would be adopted at Kyoto was logical and reasonable, based on the final outcome of the seventh COP and the adoption of the review process. As a result, Zimbabwean officials were extremely upset with what they considered the politicization of the elephant protection issue. However, the outcome was clearly not entirely unexpected. In preparation for the eighth COP, Zimbabwe and its SACIM partners had submitted a proposal to list the northern Atlantic herring in Appendix 1. The proposal, which was weak in supporting data, was withdrawn after discussion allowed Zimbabwe to make its point that CITES listings were increasingly being used for political purposes and not grounded in any scientific criteria. The herring was chosen because it is an important commercial commodity for many European countries, just as, Zimbabwe argues, the elephant is for many African range states.

It has been observed that the charge of 'politicization' of conservation decisions within CITES is largely levelled against arguments and proposals that are counter to the interests and aims of the party so charging, and that it has been levelled against proponents and opponents of sustainable use alike. It is a somewhat disingenuous charge, in that the CITES regime – like all international treaty organizations – is inherently political. Parties use politics within CITES, as in any political organization, to promote changes that will serve the Parties' own self-interested visions for the future. These changes involve both the structure of the regime itself, as well as the domain over which the regime exerts regulatory and normative influence. In the CITES context, however, a charge of 'politicization' implies that the decision in question has been taken without regard for the scientific evidence. This implication, and its converse – that a decision based on scientific evidence is somehow devoid of political considerations – are not merely naive, but manipulative. As with most great environmental debates of our age, there is often sound scientific evidence available to support any reasonable position. Such decisions, however, are taken not only in the context of scientific research but of political and economic realities as well. Nonetheless, derogatory charges of 'politicization' began to multiply at the Kyoto COP, and in the coverage in the popular press and according to some NGOs, this seemed to herald the discrediting and potential downfall of the CITES regime itself. TRAFFIC's analysis of the eighth COP included the following assessment: 'Many conference decisions were made without regard for scientific data... with the results reflecting political expediency rather than practical conservation' (Hemley, 1992, p 1).

It was at the eighth COP in Kyoto that a number of underlying tensions in CITES were brought to a head and the dispute over the Appendix I listing of the African elephant became the trigger for a larger debate over the appropriate environmental conservation paradigm to be employed by the CITES regime: sustainable use or preservation. The southern Africans were vocal advocates of the view that in many cases wildlife can only be conserved by exploiting it for economic gain. A SACIM proposal calling for CITES to recognize the benefits of trade for wildlife conservation resulted in Resolution Conf 8.3, which acknowledges 'that commercial trade may be beneficial to the conservation of species...when carried out at levels that are not detrimental to the survival of the species in question'. The importance of this resolution was overshadowed by SACIM's defeat on the downlisting proposal, but it served as a precursor to bigger changes later on.

The southern Africans stressed that the economic profit that can be gleaned from the use of wildlife gives value to that wildlife and thus provides a genuine motive for people to manage and conserve the resource. Such use might range from eco-tourism to selling hunting rights to marketing products from culled wild or ranched animals. Some Americans, Europeans and members of the environmental NGO community, however, were suspicious of the appeal to the sustainable-use paradigm within the CITES context. They believed it was being used, in some cases, as a front for an economically driven pro-trade position, regardless of the effect of such policies on wildlife conservation.

Pushing the sustainable-use approach in CITES was one way Zimbabwe began to link its economic interests in trading ivory to an environmental idea that was gaining popularity elsewhere. For example, the 1992 United Nations Conference on Environment and Development (UNCED), held in Brazil three months after the eighth COP, insisted that environmental protection and economic development must go hand in hand, if either is to be achieved. Sustainable use is a fundamental principle of the Convention on Biological Diversity, which was signed at UNCED by some 150 countries. Southern African delegations openly wondered whether that philosophy was compatible with the CITES regime, which they saw as taking a fundamentally different approach.

CRISIS AS A PRELUDE TO CHANGE

The Kyoto Conference initiated what can only be called a constitutional crisis for CITES. At Kyoto, CITES began to focus its attention on species which represented large-scale commercial industries for range states and consumer states alike. The African elephant was only one of these species; others included the bluefin tuna and several tropical timber species. In each case, member states with important economic interests in the species argued that it should be kept off the CITES appendices. When its downlisting proposal was rejected at Kyoto, Zimbabwe threatened to withdraw from CITES. Instead, it stayed and drafted a document known as CITES II which was floated informally to several Parties. The report, a rather radical denunciation of CITES, was drafted by Zimbabwe's Scientific Authority. Its stated purpose was to transform 'the present CITES to a new convention which rectifies some of the perceived defects and is more closely aligned with modern conservation concepts'. These 'perceived defects' included claims that CITES' protection had not measurably improved the status of any listed species; that CITES is founded on outdated conservation principles that are inconsistent with the goal of sustainable development; and that it is an imperialist treaty, codifying 'the entrenched dominance of Western importing states... [which] is a source of political irritation to developing countries'.

In responding to a questionnaire about CITES in October 1992 (about the time CITES II was being floated), a Zimbabwean CITES representative wrote:

Q. What have been the advantages and disadvantages for Zimbabwe in participating in CITES? On balance, which dominate?

a) *Advantages: It is very difficult to think of any. In a nihilistic sense we have needed to be members of CITES to ensure that some wildlife products (eg crocodile skins) find their best markets. Certainly, there have been no conservation advantages.*
b) *Disadvantages: They are legion. Without considering the obvious case of listing elephants on Appendix I (to which it could be argued that we would be no worse off as non-Parties), there are a number of cases where the bureaucracy surrounding CITES business seriously prejudices our attempts to promote wildlife as a general form of land use in Zimbabwe...*
c) *Without a doubt, the disadvantages of CITES totally outweigh any benefits attached to being Parties. It would be accurate to state that we are forced to remain in the CITES forum if only to protect our interests.*[1]

This is clearly an unhappy and disillusioned view of CITES, but it does indicate the beginning of a shift from total rejection to a decision to remain within the organization and to attempt to change it from the inside.

THE FORT LAUDERDALE COP

Only two proposals relating to elephants were brought to the ninth COP, held in 1994 in Fort Lauderdale. South Africa proposed downlisting its elephants to Appendix II, to permit trade in hides, hair, and

---

[1] Anonymous Zimbabwe official, written response to questionnaire by author, October 1992

meat only, with a continued ban on ivory trade; and Sudan asked for a one-time downlisting of its elephant population to allow it to sell its ivory stockpile. Zimbabwe and the other SACIM countries did not offer any downlisting proposals, probably because they believed that there was no chance of success at a meeting held in the US. The South African proposal was opposed by virtually all other African range states, including Zimbabwe and its SACIM partners. The US delegation, while finding the proposal scientifically sound, said it could not support it in the face of range state opposition. South Africa later withdrew its proposal when it became obvious that it could not garner sufficient votes for its adoption.

It is critical to understand why Zimbabwe did not support the South Africans in their effort. First, they believed that the South African proposal did not go far enough and therefore did not deserve their support. Moreover, they feared that the adoption of South Africa's proposal and a trade in elephant hide could cause the ivory trade to be banned permanently. Finally, and perhaps most importantly, Zimbabwe was changing its strategies for dealing with CITES. Zimbabwe had begun to combine its protests and denunciation of CITES with increasingly successful efforts to increase its power within CITES and to change the way the organization works. As one official put it:

> *[We have discovered that it is] better to work on CITES from within. It doesn't end with elephants; once you are an outsider you have no input or involvement...... We realize we will benefit from staying in [CITES], and now we are hosting [the next COP].*[2]

THE HARARE COP

The 1997 Harare COP represented a swing back of the pendulum from the period of the Kyoto crisis towards a less polarized approach both to the African elephant question and to the general issue of the place of sustainable use in conservation. Although the discussions at the conference were often heated, in the end the COP accepted the proposals of Zimbabwe, Botswana and Namibia to downlist their elephant populations from Appendix I to Appendix II. The conditions of the downlisting are:

---

2   Nhema, Claudius, Counsellor, Embassy of Zimbabwe. Interview with author, October 1994

- only these populations will be downlisted and all other African elephant populations will remain on Appendix I;
- the ban on the ivory trade continues until at least 21 months after the end of the tenth COP (March 1999 at the earliest) and the conditions are certified to have been met;
- Zimbabwe, Botswana and Namibia agree to withdraw their reservations to the Appendix I listing of the African elephant;
- a series of trade control and management mechanisms are certified by the CITES Standing Committee and Secretariat to be in place in the exporting countries;
- international law enforcement cooperation and reporting and monitoring systems have to be put in place; and
- the trade must begin with an experimental quota for legally held and registered ivory stocks, which can be shipped in one shipment each from Zimbabwe, Botswana and Namibia to Japan only.

Continued trade in ivory after the one time shipments would require new proposals to be submitted and approved at the next COP in early 2000.

Unlike at the Fort Lauderdale COP, where range state opposition to the South African downlisting proposal was responsible to a large extent for its failure, the range states were united in support of the downlisting proposal from Zimbabwe, Botswana and Namibia. This regional consensus was instrumental in garnering the support of other states and international wildlife organizations for the proposal.

An additional outcome of the Harare COP was the establishment of a new mechanism to generate funds for elephant conservation. Range states will declare their government-held ivory stocks, which will then be made available for a one-time purchase for noncommercial purposes. The revenue from these sales will be managed for elephant conservation through conservation trust funds. The issue of whether such stockpiles would be required to be destroyed after the purchase is unresolved and potentially highly controversial, as the release of large new quantities of ivory may pose security risks for range and consumer states alike.

ZIMBABWE'S INFLUENCE ON CITES

Due to the high visibility and political volatility of the elephant downlisting issue at the eighth COP in Kyoto, it appeared that the southern African elephant range states left that meeting very much as losers, with only the sop of a much compromised resolution on the potential benefits of commercial trade. However, on a number of less obvious issues, these countries saw several of their resolutions accepted by the Parties. Some of these, including new listing criteria for species, had the potential to influence the future of CITES in more fundamental and permanent ways than any downlisting of the African elephant.

One of the most important documents to come out of the 1994 Fort Lauderdale COP was Resolution Conf 9.24: Criteria for Amendment of Appendix I and II which, adopted against the initial wishes of a number of economically powerful Parties including the US, laid out new procedures for listing, downlisting and removing species on CITES appendices. The adoption of the new criteria can be traced to the previous COP, where Zimbabwe introduced a new set of listing criteria, which were referred to as the 'Kyoto Criteria'. Zimbabwe's proposal was rejected by the Parties, but a resolution was adopted (Resolution Conf 8.20) which directed the Standing Committee to draft a revision of the 'Berne Criteria' for listing species. Among the new listing criteria that Zimbabwe presented at Kyoto was the requirement that the Party proposing the listing must consult with the range states prior to introducing the proposal to the Parties. This was an important theme in Zimbabwe's charges – elaborated in the CITES II document – that the convention is imperialist and dominated by wealthy wildlife consumer countries who do not take the views of the range states into account. The new criteria adopted at Fort Lauderdale refer explicitly on two separate occasions to the importance of consulting with range states, and require that details of range state management, monitoring, and conservation programmes for the species in question be provided before listing is considered by the CITES Parties.

Other new factors that must be taken into account include biological data concerning population status and trends, distribution, habitat and threats. In addition, data about utilization and trade in the species must be provided. Many observers believe that the net result of the new criteria, which are more scientific and objective than listings based on the Berne criteria, will be to make it more difficult to list species on Appendix I, and easier to downlist from Appendix I to Appendix II, which would clearly be of benefit to Zimbabwe and other pro-trade countries.

Southern Africa was also instrumental in changing the committee and power structure of the CITES regime itself so as to accord more weight to the input of developing countries and range states. For example, the CITES Standing Committee is the most powerful constituent organ of the convention. Prior to the Fort Lauderdale COP, the membership of the Standing Committee was composed of one Party from each of six geographical regions: Africa, Asia, Europe, North America, Oceania, and South and Central America and the Caribbean. The host countries of the previous and upcoming COPs are also Standing Committee members. This formula gave disproportionally high representation to regions with fewer Parties – such as North America, with three, and Oceania, with four – at the expense of the larger Asian and African regions. Zimbabwe certainly viewed the Standing Committee's composition as further evidence of the 'imperialist' nature of CITES. At Fort Lauderdale, Malawi introduced a proposal to more fairly reapportion representation on the Standing Committee. The proposal met with great support from the majority of developing countries and was adopted by the Parties. The new Standing Committee provides for approximately one representative for every fifteen Parties, allowing Africa three, Asia two, Europe two, South and Central America two, and North America and Oceania one each. The African representatives are Namibia, Senegal, and Sudan.

In addition to composition changes, the Fort Lauderdale COP also resulted in an unexpected change in the Standing Committee chairmanship. When New Zealand's term ended, the position did not go to the UK, as expected, but to Japan. Although Japan is a large industrialized country and a major consumer of many CITES-listed species and products, it tends not to share the views of some other consumer states, such as the US and many EU states. Instead, it supports a pro-use paradigm, particularly regarding elephant ivory, of which Japan was the world's largest importer before the 1989 trade ban. Leading up to the Fort Lauderdale COP, Japan had taken a record number of reservations to listed species in CITES, and it had received a less than favourable assessment by the CITES secretariat for its record in implementing CITES regulations at the national level. In 1994, internal changes and global dynamics were converging to change Japan's approach to, and role in, CITES. Japan's agreement with southern African range states on the ivory trade issue, as well as on broader questions about the role of CITES, served Zimbabwe well as Japan came to assume a leadership role in the organization. The Standing Committee changes that came out of Fort Lauderdale did much to bring CITES more into line with Zimbabwe's 'pro-trade' and 'anti-imperialist' visions.

The ninth COP provided further evidence that southern African positions were supported by other industrialized nations. Document

9.18, entitled *How to Improve the Effectiveness of the Convention*, was a very general call for a study of the regime's overall effectiveness by an independent consultant, with findings to be reported to the next COP. The document was adopted in committee by a vote of 62 to 4, with no opposition voiced in plenary session. The proposal was widely believed to be a direct descendent of the CITES II document circulated by Zimbabwe. The widespread support for its adoption was, at least in part, the result of four years of behind-the-scenes coalition building on its part. The US was among the four opposing votes because of concerns regarding the measurement of regime effectiveness, elaborated below, and because it felt the money could be better spent on improving CITES implementation. The effectiveness proposal is not radical or critical in tone, as was the CITES II document, but several themes from CITES II appear in it. One is CITES' age. CITES II argued that the regime embodied outdated principles and goals that have since been discredited. The new proposal states, with a similar implication, that 'during that period [since 1973] the number of conservation conventions has multiplied many times', thus supporting the need for review.

CITES II claimed that 'there is only one measure [of the success of CITES] and that is the extent to which species populations have increased as a result of CITES'. Along the same lines, the effectiveness proposal directs the consultant doing the study of CITES to provide information about the extent to which the conservation status of a representative selection of species listed in each of the three appendices of CITES has changed, and the extent to which the change can be attributed to the application of CITES.

It is extremely difficult to assess the effectiveness of a regime. Effectiveness may take the form of increasing public awareness, education, strengthening domestic conservation structures and procedures through economic assistance or technical support, decreasing demand in consumer countries, or deterring potential behaviour that is prohibited by the regime. As international trade usually represents only a tiny fraction of the pressure on threatened and endangered species, a narrow assessment, such as that outlined in the effectiveness proposal, was always likely to conclude that CITES is doing little or nothing for its listed species. This could theoretically make it easier for Zimbabwe and others to steer CITES away from listing species, particularly on Appendix I, and toward a pro-use stance. Although this study had the potential to influence CITES more strongly in the long run than any other outcome of the Fort Lauderdale COP, the results of the study presented in Harare did not meet anyone's expectations or needs, and, overshadowed by the African elephant issue, the study anti-climactically played a minor role in the proceedings.

Finally, at Fort Lauderdale, Zimbabwe was designated the host of the tenth COP, a role which it intended to use to 'educate delegations about the real situation'.[3] Based on the outcomes of the Harare COP, it can be argued that Zimbabwe was successful.

It is difficult to determine whether these changes to CITES are a direct result of Zimbabwe's strategies and actions, or whether Zimbabwe is helping to facilitate trends originating elsewhere. Such larger trends may emerge from within an informal coalition of like-minded countries, or may represent underlying outside forces such as the sustainable development rhetoric that came out of the 1992 UNCED. Most probably, the changes within CITES result from all those dynamics simultaneously. In any case, Zimbabwe has become increasingly effective in both facilitating change and using it to forward its national interests within the CITES regime.

THE DOMESTIC DIMENSION

To understand Zimbabwe's shift from emphasizing its opposition to the Appendix I listing of the African elephant to its more recent efforts to promote a sustainable use paradigm within CITES, it is essential to grasp the domestic politics of wildlife conservation in Zimbabwe. Zimbabwe's position in CITES has been shaped by the emergence of a coalition of groups within the country favouring sustainable use. Before Zimbabwe gained independence in 1980, wildlife conservation was largely under the direct control of the colonial state. One change to this came in 1975 when rights to use wildlife were conferred on white landowners. This enabled them to profit from wildlife ranching and safari hunting. But it was still the case that the black majority had no rights over wildlife and that they were often hostile to the state's conservation policies.

After independence and the advent of black majority rule, there was a growing awareness that this situation could not continue. Although Zimbabwe's conservationist NGOs were predominantly white, they recognized that the cooperation of black communal farmers was essential to the success of conservation policies. In order to secure this cooperation, officials within the Department of National

---

3   Nhema, Claudius, Counsellor, Embassy of Zimbabwe. Interview with author, October 1994

Parks and Wildlife Management (DNPWLM) and the NGOs in the environment and development sectors agreed that these farmers must have a vested interest in the value of the wildlife on their land. Indigenous communities should be given the same sort of rights that white farmers had acquired in the 1970s. This resulted in the development of the Communal Areas Management Programme for Indigenous Resources (CAMPFIRE). This programme got underway in 1988 and it resulted in bringing more Africans into the field of wildlife conservation. It gave rural communities a greater degree of control over wildlife and it provided hard evidence that wildlife conservation can provide revenue and jobs to communities, whether through tourism, selling select hunting rights, or selective exploitation for food and other products. The DNPWLM says that CAMPFIRE has been successful in many communities where it has been implemented. For example, the Nyaminyami district, which was the first to implement CAMPFIRE, raised $458,000 in 1992 by selling hunting permits, meat and hides. This money has been put towards community healthcare, schools, water systems, and arming wildlife protection rangers.

There have been tensions within CAMPFIRE, however. There are disagreements about the proper distribution of revenues between district councils on the one hand, and local communities on the other. Nevertheless, the coalition supporting the sustainable use of wildlife is a broad one in Zimbabwe. In contrast to the US and Europe, where NGOs often play the role of opposition to government, it incorporates the positions of both government officials and most NGOs. It also includes rural communities who benefit from CAMPFIRE schemes and white landowners who engage in wildlife ranching.

The existence of this coalition provided a powerful impetus for Zimbabwe to pursue changes within CITES. The policies of CITES have not generally been favourable to the development of sustainable use. Yet, Zimbabwe has come to feel it must be part of CITES if it is ever to earn the goodwill of the international community and to secure the important development funding and trade benefits that come with it. Hence, Zimbabwean officials have sought to change CITES policies towards ones more compatible with Zimbabwe's practice of promoting sustainable use. Zimbabwe's advocacy of sustainable use has met with scepticism from some within CITES. The success of CAMPFIRE has been questioned and the problems with district councils highlighted. International NGOs have pointed to Zimbabwe's failure to control rhino poaching. It is suggested that this does not bode well for the reopening of the ivory trade. There have also been allegations of corruption and the distortion of data. While not all these charges are well founded, they do indicate the difficulties of determining the validity of various conservation strategies, including sustainable use, within CITES.

CONCLUSIONS

Zimbabwe's behaviour as a CITES member points to two conclusions. First it shows that CITES membership has made a difference in policy decisions taken by Zimbabwe's leaders. The decision to adhere to the ivory trade ban, while working to overturn it, was in accord with CITES rules, and in direct contrast to the state's previous conception of its national interest.

Secondly, this case study shows that Zimbabwe, as a Party to CITES, has learned to use the regime and has been instrumental in bringing about – and taking advantage of – profound changes in the intent, the structure, and the power relationships embodied in the CITES regime. Zimbabwe's relationship to the regime has evolved from being the recipient of regime dictates to being the creator of some of those dictates, often in a form that serves the state's self-calculated national interest. It also shows that international cooperation efforts to protect natural resources are re-defining traditional notions of wealth and power. For many species, the CITES range states tend to be poor in traditional economic power terms, but rich in the resources in question. As value is assigned to newly acknowledged forms of wealth (eg endangered species), countries rich in these resources are learning to derive power through the leverage they provide in the context of regimes like CITES.

CITES, which was created to regulate international trade in threatened wildlife species, is now tackling the issues of habitat conversion and wildlife management. In part, it is increased scientific knowledge which has led species protection efforts in this direction. However, it is also the result of Parties learning to forward their individual economic interests through the regime. As Zimbabwe has learned how to influence the CITES regime it has empowered itself, giving it more input in important multilateral issues than it would have had in the absence of the regime. Other range states within CITES have also fundamentally improved their ability and willingness to speak out. The regime–state relationship is neither static nor unidirectional. Rather, the relationship is dynamic and reciprocal; it develops over time as states learn to work within the system to advance their interests.

REFERENCES

Hemley, G (1992) 'CITES 1992: Endangered Treaty', *TRAFFIC USA*, 11 August

Ricciuti, E (1993) 'The elephant wars', *Wildlife Conservation*, March–April

Thomsen, J (1989) 'Conserving the African Elephant: CITES fails – US Acts', *TRAFFIC USA*, 9 January

Part IV

THE FUTURE OF CITES

Chapter 11

CITES and the CBD

R B Martin

INTRODUCTION

The Secretariats of the Convention on International Trade in Endangered Species of Wild Fauna and Flora and the Convention on Biological Diversity signed a Memorandum of Understanding on 23rd March 1996. The articles of neither treaty require this but it was a decision of the Parties to both conventions to move in this direction.[1] The memorandum provides for institutional cooperation between the Secretariats, exchange of information, coordination of work programmes and joint conservation action. At this stage, the understanding goes no further than this limited cooperation between the Secretariats and is careful not to commit the Parties beyond the decisions that they have already adopted. Nevertheless, the memorandum states: 'The secretariats will consult their Contracting Parties with a view to encouraging... effective conservation and promoting the sustainability of any use of wildlife as a part of the biological diversity of our planet...'. This is perhaps the first time that CITES has formally used the language of sustainable use and biological diversity. This language is central to the CBD. So, the memorandum provides some grounds for thinking that CITES and the CBD may be moving closer together.

While a process of convergence may have been initiated, it remains true that the two conventions have very different origins and

---

[1] The CITES Parties endorsed the Strategic Plan of their Secretariat at the ninth COP held in Fort Lauderdale, US, in 1994. This included, *inter alia*, a statement to the effect that special attention should be given to the Secretariats of other environmental conventions including the CBD. The Parties to the CBD adopted a Decision (II/13) at their second COP, held in Jakarta, Indonesia, in 1995, requesting their Secretariat to coordinate with the Secretariats of other conventions.

structures. CITES came into force on 1st July 1975 following a three-week plenipotentiary conference in Washington DC in 1973.[2] At the time, there was very little knowledge of the nature and magnitude of international trade in specimens of wild species. Nevertheless, there was a strong feeling that international trade was a significant cause of species decline and the articles of the treaty are replete with the restrictive conservation language of the 1970s.

The CBD is a much more recent treaty that emerged out of the United Nations Conference on Environment and Development (UNCED) held in Rio de Janeiro in 1992. It was conceived at a time when 'sustainable development' had become a key notion in conservation. The first meeting of the COP of the CBD was held in the Bahamas in 1994 and this has been followed with a meeting every year since then. The CBD has also established a Subsidiary Body on Scientific, Technical and Technological Advice (SBSTTA) which meets annually, with its agenda set by the COP.[3] In addition to this, regional caucuses of the Parties to the CBD also meet prior to the COP. Compared to CITES, which meets once every two to three years, the CBD is far more complex. This chapter assesses the relationship between CITES and the CBD. It begins by considering what common ground they share. It then examines some of the differences in the ways in which they pursue their aims. Finally, a case is put for subsuming CITES under the CBD.

COMMON GROUND?

In discussions of the common ground between CITES and the CBD three questions are frequently raised. Do the conventions agree on what is to be conserved? How do the conventions understand the threats to biodiversity? What is their view of the place of sustainable use in conservation? On the first question, there might appear to be a difference. CITES, as its name indicates, is concerned with conserving

---

2   It is of interest to note that many Parties still refer to CITES as the 'Washington Convention' and steadfastly refuse to use the soubriquet 'CITES'. This practice is not simply a quaint throwback. Because of the close resemblance between CITES and the United States Endangered Species Act, many states feel that CITES is a Western creation.

3   The SBSTTA appears to be undergoing a minor 'identity crisis'. As the Chairman remarked at its 2nd meeting in September 1996, SBSTTA should be neither a 'mini-conference of the Parties' nor 'a drafting group' but rather an advisory group on scientific and technical policy.

species, while the primary focus of the CBD seems to be on ecosystems. This difference has been emphasized by those who are anxious not to see CITES lost in the larger, all-inclusive purview of the CBD. However, the distinction is not a very significant one in this context, since ecosystems are made up of species, and CITES does, in any case, recognize the role of species in ecosystems.[4]

With regard to the threats to biodiversity, the CBD has carried out a preliminary identification of the proximate threats to biological diversity and lists the following:[5]

- habitat destruction or deterioration;
- introduced species;
- overharvest or overkill of wild species;
- pollution; and
- climate change.

Of these, there seems to be general agreement that habitat loss is probably the greatest threat to biological diversity and ecosystems. In the same CBD document, the ultimate causes of these threats are identified as:

- land tenure;
- population change;
- cost-benefit imbalances;
- cultural factors;
- misdirected economic factors; and
- national policy failure.

It is significant that international trade in wildlife products is not identified as one of the significant causes of the loss of biological diversity, despite this being the sole *raison d'être* for CITES. CITES concentrates on just one threat, whereas the CBD deals with many.

---

[4]   Article IV.3, in dealing with regulation of trade in specimens of species included in Appendix II, states that 'Whenever a Scientific Authority determines that the export of specimens of any species should be limited in order to maintain the species throughout its range at a level consistent with its role in the ecosystems in which it occurs and well above that level at which the species might become eligible for inclusion in Appendix I...' And in the criteria for listing species on the Appendices of CITES which were adopted at the ninth COP, considerable weight is given to the possible impact of the exploitation of a species on other components of the ecosystem.

[5]   CBD Secretariat's note to the Subsidiary Body on Scientific, Technical and Technological Advice (Document UNEP/CBD/SBSTTA/2/3).

The one threat dealt with by CITES appears not to be given any weight by the CBD.

On the third question, the CBD is strongly committed to sustainable use. This is evident from the first article of the treaty where the first two objectives mentioned are 'the conservation of biological diversity' and 'the sustainable use of its components'. CITES' commitment to sustainable use is a more contentious matter. The declaration by the Secretariats of the two conventions in which there is a commitment to sustainable use has already been mentioned, but there is no mention of sustainable use in the preamble or articles of CITES. Indeed, the very brief preamble to CITES limits itself to a statement that 'international cooperation is essential for the *protection* of certain species of wild fauna and flora against *overexploitation* through international trade' (emphasis added). The subsequent articles present trade as being neutral or detrimental to species survival. It was only in 1992 that the Parties to CITES recognized that trade might be beneficial for conservation.[6] Moreover, in the recent study on the effectiveness of CITES the consultants noted that a number of Parties did not see sustainable use as being the central theme of CITES. It caused the consultants to recommend that 'The issue of sustainable utilization and its relation to CITES should be addressed in an interpretative Resolution by the Conference of the Parties as a matter of priority' (*Environmental Resources Management*, 1996, Recommendation 3C). On the other hand, in regulating trade in Appendix II species CITES is implicitly concerned with sustainable use. Moreover, a recent publication by the Humane Society of the United States (HSUS) claims that the concept of sustainable use is enshrined in CITES (Wold, 1995). Stated or unstated, it is clear that CITES, in attempting to regulate international trade, is pursuing the objective of the sustainable-use of species and ecosystems.

With some qualifications, then, CITES and the CBD have the same goals, and they both accept sustainable use as a valuable conservation tool. But, while the CBD acknowledges a wide range of threats to conservation, CITES is concerned with only one: that emanating from the international trade in wildlife. Thus, CITES' activities can be regarded as contained within a small subset of the activities of the CBD. The HSUS' claim that the CBD should defer to existing institutions (Wold, 1995, p 6) appears like an attempt to get the tail to wag the dog.

---

6   See Resolution Conf 8.24 adopted at the eighth COP, in Kyoto, Japan, 1992.

DIFFERENCES IN APPROACH

While both CITES and the CBD are committed to the conservation of wild species of flora and fauna, the means by which they seek to realize this goal are very different. Eight specific differences can be noted.

1   CITES operates in a vacuum, taking no account of human or economic considerations. This is very much out of step with the growing global realization that conservation and sustainable use in the developing countries can only be achieved with the participation of local peoples. In contrast, the CBD recognizes that economic and social development and poverty eradication are the first and overriding priorities of developing countries (Preamble, paragraph 19) and takes into account the interests of local communities in conserving and using biological diversity sustainably (Article 8(j)).

2   CITES focuses on the global level. If a species is considered endangered at the global level then, regardless of its status in any individual country, trade may be banned.[7] Moreover, through this approach CITES effectively treats the wealth of natural resources in individual countries as part of the global commons or, as it is sometimes put, as part of the Common Heritage of Humankind. This provides the justification for some states to impose their conservation solutions on others. The CBD has so far avoided such pitfalls. It focuses on nation states, stipulating that 'States are responsible for conserving their biological diversity and for using their biological resources in a sustainable manner' (Preamble, paragraph 5). It recognizes that 'States have sovereign rights over their own biological resources' (Preamble, paragraph 4).

3   CITES does not consider the possibility that trade may have benefits for species, ecosystems or people. The assumption is that international trade is the greatest threat to species survival. The articles of CITES embody a consistently negative view of trade. Article III 3 (c) states: 'An import permit shall only be granted when ... the specimen is not to be used for primarily commercial purposes.' Article IV 2 (a) states 'The export of any specimen of a species included in Appendix II ... [shall only take place when] such export will not be detrimental to the survival of that species'. In a document endorsed by the Secretariat of CITES, the HSUS

---

[7]   The inherent inconsistency in this situation seems to escape the Parties to CITES: if a species is secure in *any* state, then its global status must also be seen as secure.

states that 'international commercial uses of biological diversity usually lead to significant population declines for species' (Wold, 1993, p 12). This is a sweeping claim, with plenty of examples to disprove it (Crocodile Specialist Group, 1996). The HSUS go on to say that '[r]esearch and cooperation should not focus on developing new uses of biological diversity' (Wold, 1993, p 12). The assumption is that things are bad enough as it is. In contrast, the CBD recognizes, in the preamble to the convention, the full range of values of biological diversity, including economic values; and, as has already been noted, it has not identified international trade as a threat to biological diversity.

4   CITES establishes an inflexible link between a species' perceived biological status and the manner in which it can be used. In this respect it follows the US Endangered Species Act: it has no proportional controls. If a species is listed on Appendix I, trade is automatically banned, even if it would benefit the survival of the species. All other species (including those listed on Appendix II) can be used unsustainably without inputs into their improved management. The fact that CITES has adopted improved criteria for listing species on its appendices is irrelevant. The flaw lies in the rigid adherence to a binary system where either all is prohibited or all is permitted. So far, the CBD has avoided this type of inflexible linkage between the biological status of a species and the manner of its use. The CBD provides for annexes, which are binding on all parties, and protocols that are only binding on signatories to any particular protocol. But, at this stage, the CBD has not adopted any annexes of practical significance to trade or sustainable use and it is as well that the Parties proceed cautiously down this route in order to avoid the prejudicial and unworkable systems that characterize CITES.

5   CITES sees itself as a science-based convention, but the demands from some Parties for scientific information, which is expensive to obtain and not essential to the sustainable use of a species, are frequently unreasonable. There is an air of unreality about such demands. The HSUS holds that no intended use should proceed without extensive biological information concerning the species being used, and that without such information one cannot be sure that use will be sustainable (Wold, 1993). This is to overlook the fact that the majority of uses are already taking place without scientific certainty. While the CBD sees the importance of science in contributing to the improved understanding of biological diversity (Preamble, paragraph 7), it has not made unreasonable scientific

demands and it appears ready to adopt a more holistic scientific approach.

6   The command and control structure of CITES leads to a preoccupation with law enforcement. If the treaty appears not to be improving the status of species, then the diagnosis is that the treaty is not being adequately enforced by its Parties. However, the costs and social implications of the proposed law enforcement measures ought to cause Parties to question the appropriateness of the treaty. In contrast, the CBD requires Parties to 'adopt economically and socially sound measures that act as incentives for the conservation and sustainable use of components of biological diversity' (Article 11).

7   In making decisions which affect other Parties in CITES, no state carries any financial responsibility for the costs which it may be placing on another state and neither are there any funds forthcoming under the treaty to assist states with such conservation costs. The CBD provides funds to support *in situ* conservation and measures aimed at sustainable use (Article 8(m)).

8   Unlike CITES, where the development of the treaty was primarily in the hands of the developed nations, the CBD has included developing country perspectives from the outset. The political alliances in the CBD are stronger than in CITES and, while this frequently leads to frustrations on the part of some Parties with the pace of progress in the technical realm, it should lead finally to solutions which are workable and acceptable to all Parties.

Taken together, these eight contrasts show that the CBD adopts a more balanced, realistic and comprehensive approach to conservation than does CITES.

SUBSUMING CITES UNDER THE CBD

In the recently completed study on how to improve the effectiveness of CITES, the consultants state:

> As the 1992 Convention on Biological Diversity (CBD) is the broadest and most politically important global conservation convention... CITES should develop more effective working relations with the CBD, as a means of setting priorities for species management and for funding special projects (*Environmental Resources Management*, 1996).

The Crocodile Specialist Group of the IUCN Species Survival Commission, in contributing to the CITES Review Process in 1996, suggested that:

> *It would be appropriate for CITES to link formally to the CBD such that critically endangered species on Appendix I, which are not well supported by CITES funding mechanisms, should become priorities for funding under the CBD* (Crocodile Specialist Group, 1996).

There would seem to be a convergence in thinking from many directions that there is a need for CITES to move closer to the CBD. The memorandum mentioned in the introduction to this chapter is itself one sign of this. The question is: 'How much closer?' In the light of the fact that the goals of CITES coincide with those of the CBD, and that the comparison between the two conventions favours the latter, there is a strong case for subsuming CITES within the CBD. It should not be too surprising that the CBD should provide a better framework for conservation. It has had the advantage of twenty years of progress in conservation thinking and practice. Two aspects of this progress are particularly important. Firstly, there is the greater flexibility offered by the CBD. The hard linkages between trade and listing on the appendices established by CITES have been shown not to work. Professor Harry Messel, who was at the inaugural CITES conference in Washington in 1973 and played a part in the drafting of the treaty, has remarked that, at that time, the state of knowledge on international trade and the means by which it might be controlled was extremely limited.[8] When asked if, with knowledge gained in hindsight through participation in many CITES meetings and his experience in the Crocodile Specialist Group of IUCN, he would approach the problems differently, he responded positively that he most certainly would. In his view, the system of appendices was unnecessary and the treaty would operate more effectively through a quota system such as is, in fact, practiced in CITES for the trade in crocodiles.

Secondly, in the last two decades, conservationists have come to recognize the important influence of social, economic, and political factors on the success of conservation. The CBD possesses all the ingredients for a holistic approach to conservation and sustainable use. It is a force for the decentralization and devolution of responsibilities to local communities, both in the developed and developing world. Its

---

8   Personal communication, 1992.

recognition of the need for incentives for people and the placing of economic value on wild resources puts it in the category of 'conservation with a human face' (Bell, 1987), as opposed to the 'command and control' regime of CITES. Thus, given the inclusion of CITES functions within the purview of the CBD and taking into account the considerable strain on all Parties to keep up with the attendance and preparation demands for international meetings (those attending CBD meeting are often the same as those at CITES), it would appear very desirable to subsume CITES under the CBD. The argument that it is very difficult to manoeuvre existing institutions is not impressive. Several treaties on conservation and sustainable use have come and gone since the turn of the century indicating that there is nothing immortal about any treaty. Given the speed with which CBD was ratified, it seems that with political consensus things can be done in a reasonable time. Both the CBD and CITES are under the control of their contracting Parties and, at the end of the day, it is their will which will determine the outcome.

REFERENCES

Bell, R (1987) 'Conservation with a human face: conflict and reconciliation on African land use planning', in Anderson, D and Grove, R (eds) *Conservation in Africa – People, Policies and Practice*, Cambridge University Press, Cambridge

Crocodile Specialist Group (1996) *CITES Review Process: The Lesson from Crocodilians*, CITES Secretariat, Lausanne

Environmental Resources Management (1996) *Study on How to Improve the Effectiveness of CITES*, Report to the Standing Committee of CITES, Lausanne

Wold, C (1995) *The Biodiversity Convention and Existing International Agreements: Opportunity for Synergy*, Humane Society of the United States and Humane Society International, Washington DC

Chapter 12

Developing CITES: Making The Convention Work for All of the Parties

Timothy Swanson

INTRODUCTION

Wildlife trade regulation can be seen from very different perspectives. Some people would like to see it as a first step toward the containment of wildlife utilization, and the prevention of the decline of individual species in other countries. Others would like to see it as a licence to use the wildlife trade for the purposes of social development, and they may view other objectives as infringements upon national sovereignty. Is there any prospect for the regulation of the wildlife trade to accomplish a broad set of objectives consonant with this wide range of perspectives: the encouragement of growth and development, and the avoidance of wildlife depletion? Or will it always remain necessary to embrace one of the two polar positions: a trade proponent or a trade protectionist?

The argument of this chapter is that it is not only possible to achieve this common objective but that it is a necessary objective to pursue. International environmental law will become rationalized once it takes a shape that positively encourages trade and development down particular pathways, rather than simply banning or discouraging a few of these. The fact is that the shape that development takes in various countries does matter to others, and they are willing to pay in order to help direct that development. This is the reason why millions of residents in some developed countries give tens of millions of dollars each year to organizations whose self-proclaimed objective is the conservation of the wildlife resident in other countries (see the discussion on environmental values in World Conservation

Monitoring Centre, 1992). People do have preferences regarding the form of development that occurs in lands that are not their own, and these preferences may be transformed into a willingness to subsidize particular pathways toward development.

Do such preferences constitute unwanted interference in the decisions that rightfully belong to other peoples? Of course not, because every purchase by every person on each and every day has the same sort of impact, whether it is done through ignorance or not. Flows of funds from consumer purchases shape the world, and taking this impact into consideration is the thoughtful, rather than the neglectful, thing to do. National sovereignty can always be protected in matters of trade, because trade requires the consent of all of the parties to the commerce. In order to protect sovereignty a trade regulation mechanism should not disallow particular forms of trade; it should simply encourage those forms that meet the complex objectives of certain purchaser groups. That is, a trade mechanism should have as its objective the combination of the consumer good and the production process into a single – tradeable – package. For example, when people care not only about the good they are buying (eg ivory) but also the process which created it (eg, free range versus factory production), then it is possible to sell the good/process in a single package and charge more for it. This is the objective of a mechanism such as 'eco-labelling': it provides the consumers with the information necessary to discriminate between the various production processes underlying otherwise indistinguishable products. Then, the consumers are able to affect the development choices of producers through their informed purchases, while continuing to allow producers to choose their own development paths.

There is another reason why it is important for trade regulation to move in this direction. International trade regulation can work only if it is devised in a manner that takes both sets of perspectives into account: consumer and producer. Otherwise, one side or the other will have little incentive to adhere to the treaty, or to support its objectives. Trade regulation created only from the perspective of one side of the exchange is doomed to failure.

In this chapter I would like to demonstrate these general points within the context of a case study of CITES. In this context it will be easy to show that the initial system of international regulation had little to offer to one side of the exchange, but that for this reason, the system (to become effective) has had to evolve increasingly toward the middle ground. This middle ground lies between the wildlife producing states and the wildlife consuming states. At the same time, it is moving toward becoming an international regime that works: it will establish the means and the mechanism through which poor countries will be able to afford to conserve wildlife within their territories. This will allow them to develop down a pathway consistent with the retention of their wildlife. CITES has taken nearly thirty years to reach a juncture closer to the middle ground between the parties involved in this trade. By learning from this process, it should be possible to develop other international regulatory regimes that have the prospect of becoming effective and useful from the outset.

THE CONTROL STRUCTURE WITHIN CITES AS DRAFTED

Of the large number of international environmental conventions, CITES has probably the single most detailed control structure. It was the first international wildlife treaty to provide for both express obligations and international monitoring. Therefore, CITES represents an important step along the road toward making substantive international law with concrete impacts. The purpose of this analysis is, however, to ascertain the capability of the convention to address, as drafted, the developing world's perspective on the endangered species problem. Other authors may be consulted for a detailed analysis of the specific workings of CITES (Lyster, 1985; Wijnstekers, 1990).

CITES was signed in March, 1973 and came into force two years later. The argument of this chapter is that CITES was originally drafted with little attention to the problems of the developing countries in maintaining their species. It focused instead on the identification of species that were endangered through commercial use, and the prohibition of trade in the same. Once a species is listed on either of the CITES appendices, it becomes subject to the permit requirements of the convention. An Appendix I species may not be shipped in the absence of the issuance of an export permit by the exporting state [Article III(2)]. In addition, this permit may not be issued unless both the exporting state certifies that the export will not be detrimental to the species, and the importing state certifies (by the issuance of an import permit) that the import will not be used for commercial purposes [Article III(3)(c)]. Therefore, an Appendix I listing acts as an effective ban on the trade in those species and, even if exporters wish to continue the trade, the importing states have the duty to deny all commercial imports. An Appendix II listing, on the other hand, leaves the decision on trade control wholly to the discretion of the exporting state. That is, there is no role for the importing state, other than to ensure that an export permit has been issued for each specimen [Article IV(4)]. These permits are allowed to be issued so long as the exporting state itself certifies that the export will not be detrimental to the survival of the species within the exporting state [Article IV(2)].

The other important responsibility of the Parties is to provide annual reports to the CITES Secretariat on the amounts of trade in listed species [Article VIII(7)]. The Secretariat also sometimes acts as the intermediary between exporting and importing states, in order to confirm the authenticity of trade documents, for example.

THE NATURE OF THE EXTINCTION PROBLEM IN DEVELOPING COUNTRIES

In the past, direct human overexploitation has been the primary factor contributing to species extinctions (Diamond, 1989), and most of the documented extinctions have occurred outside of the tropical zones. In pre-historical terms, much of the megafauna of the northern temperate zones was simply hunted to extinction. This has also been a problem in regard to some oceanic species exploited by Northerners (eg the near extinction of the Blue Whale and the severe reduction of various fish species). Therefore, given this historical record, it is not surprising that those in the developed world fear the worst from wildlife exploitation. This is not always the case with regard to many of the peoples of the developing world and their wildlife. In many cases, these countries have a long history of co-existence between people and wildlife. Partly on account of this, most of the world's remaining diversity of species resides in these developing countries. However, at present and in the future, there is a very serious threat to these species looming in these same countries. Yet, it is still not the threat of over-exploitation that most endangers the species of these countries. It is the loss of the habitat on which these species rely that is likely to be the more significant factor contributing to species extinctions in these lands.

The countries with the most species are facing unprecedented development and population pressures. In these areas a doubling and redoubling of the human populations is a virtual certainty over the next 50–100 years. Kenya, for example, has a current population growth rate of about 4 per cent, which implies a doubling of the human population every 18 years. This is not exceptional; population growth rates of 3–4 per cent are the norm throughout much of Sub–Saharan Africa, Latin America and Southeast Asia.

In addition to human population growth, there is also great pressure for development in these regions. Of the 15 countries that feature prominently in terms of diversity of higher species (reptiles, birds, and mammals), none has an average annual income greater than US$2,000. In fact, most of these countries register average incomes that are among the lowest in the world, around $200–$500 *per annum*. In essence, the vast majority of the world's wealth of species lies within the borders of the poorest of nations (Swanson, 1992).

It is the pressures of human populations and development needs in these poorest of countries that currently represent the greatest threat to species. As population and development pressures continue to mount, they result in the conversion of massive quantities of previously available wildlife habitat (Wilson, 1988). The felling of tropical forests and the extension of the frontier into these regions relieves some of these pressures temporarily, but at a tremendous cost in terms of the loss of species. From the perspective of the developing world, these are the base problems of species extinction. Therefore, it is still the human factor that is extinguishing species via habitat conversions, but now the greater threat is that it will do so *indirectly*. Species are finding themselves undercut when humans convert the resources on which they rely to human uses.

The problem with CITES is that it is a global treaty built on a single perspective. A policy for the reduction of species extinction *must* address *both* of these sources of losses: overexploitation and habitat conversions. Efforts at saving species from overexploitation, where the resources of developing countries are concerned, may actually be only exercises in shifting them from the one column of the extinction ledger to the other. Since the vast majority of global wildlife exists within the developing world, a global treaty really must focus upon the needs of this sector. Over the two decades or so of operations under CITES, the Parties have become aware of the dual nature of the endangered species problem. It has been recognized within the context of the COPs that CITES must evolve to meet both prongs of the problem, and the direction of change is slowly being revealed.

THE PATH DOWN WHICH CITES MUST DEVELOP

In order to address both facets of the extinction problem, it is necessary to do two things. First, it is necessary to route funds to those countries with large numbers of endangered species, so that they might then provide habitat for these species and closely manage the exploitation of those species. Second, it is essential that there is assurance that the funds supplied are directed to the purposes indicated: the conservation of endangered species. The provision of funding alone is no solution. There must also be incentives in place to encourage the investment of the funds in species and habitat conservation. This is because terrestrial species reside almost without exception within the borders of particular states, and ultimately the question of conservation concerns the incentives provided to the range states involved.

For these purposes, a strategy of wildlife utilization by local peoples has much to commend itself as a conservationist tool. That is, using the unique characteristics of wildlife habitat to generate funding (through tourism, wildlife products trade, subsistence harvests) can itself act to conserve that same habitat. It creates revenues for species conservation and it creates incentives for the application of those funds to that purpose. People are willing to abstain from converting habitat that is valuable in its current state. It is the fact that it has this combination of capacities (ie, fund raising and incentives generating) that makes a policy of wildlife utilization so appealing. However, it is also historically apparent that unregulated wildlife utilization is a threat as well as a potential benefit. With the diffusion of new technologies and the decay of old institutions, over-exploitation has come to endanger the wildlife of developing countries as well (Luxmoore and Swanson, 1992).

The major problem with this approach is the creation of a system for the sustainable management of natural habitat and its resources. This is an expensive proposition on account of the unregimented nature of such habitat. Most developing countries are unequal to such a sophisticated and expensive, management task in the context of rapidly changing (and often deteriorating) conditions. Therefore, as a possible direction for the evolution of CITES, the regulation of trade to achieve this goal is one possible approach to the dual problem of extinction.

Another possible direction for the evolution of CITES is the movement toward captive breeding of wildlife for trade. Captive breeding is a term defined under CITES, which means the maintenance of a number of individuals as breeding stock completely segregated from the wild population. Then, under such a regime, the progeny of the breeding stock are available for trade, even when the wild variety of the same species are not. Obviously, the captive breeding policy addresses the problem of over-exploitation. If captive bred animals, (such as the fur-bearing animals: minks, otters, foxes), are provided in adequate numbers and at low enough prices, then the pressure on those same species in the wild is reduced.

However, captive breeding does nothing to address the problem of habitat conversions. In fact, to a large extent, it worsens this situation. This is because captive breeding operations are usually operated in the consumer states, where the bias is toward intensive farming methods of production. Therefore, a shift from wild harvests to captive breeding methods is largely a movement of the wealth of species from the developing countries to the developed. Once the wildlife species valued by the developed countries are established under farming regimes in the North, the natural habitats of the South are rendered truly valueless in the eyes of those living on the frontiers of these wildernesses. These people will not have any incentive to conserve the wildernesses and there is no deterrent to the law of the chainsaw (Swanson, 1992).

Another possibility is to preserve habitats as parks and protected areas, while preventing over-exploitation through captive breeding or a similar policy. This has in fact been attempted throughout much of the developing world. In the 1970s and the first part of the 1980s, vast areas of land were designated protected areas. Today, nearly 4 per cent of remaining wilderness has received some sort of protected status. There are a number of reasons why this policy is not effective. First, parks without funds are not protected areas in any sense of that term. It has been found that the level of poaching in African game parks is directly related to the level of expenditure on the parks. A figure of nearly $200 per square kilometre was necessary for effective prevention. Many of the parks in the developing world exist almost wholly on documents alone. There is no funding available for such extravagances in countries where infant mortality rates are high and literacy rates low. Most of these are paper parks only.

Secondly, the prevalence of poaching points out the contradictory nature of the policy. None of these natural habitats evolved without a human component within them. The local peoples have usually used the lands and appurtenant resources over the centuries. These attempts at halting all utilization, which are as old as the colonial legacy in these regions, introduce conflicts between local peoples and local wildlife which were not there historically (Marks, 1984).

Finally, all of these problems could be sorted out if substantial sums of money were forthcoming from the international community for their resolution. However, in general, there are many expressions of good intentions but little solid support from this sphere. The level of financial support pledged at the time of the ban of the ivory trade in 1989 never came to fruition. The developed world, through the mechanism of the Global Environmental Facility (GEF), pledged $200 million per year toward biodiversity-related projects over the first half dozen years of the GEF. The scale of this funding is large compared to past contributions, but it is as nothing compared to the value of the resources at stake. To put this amount into context, $200 million is roughly the amount that the developed world pays the producer states for raw reptile skins alone each and every year.

This quick review of the current options indicates that the route that holds most promise for addressing the developing world's species conservation problems is that of *controlled* wildlife utilization. The use of the wildlife trade to generate incentives for species conservation in the developing world would be effective if the developed world routed its custom only to those countries that demonstrated sustainable management of their natural habitat. This form of environmental conditionality would punish unsustainable exploitation, while encouraging sustainable exploitation and investment in habitat, and while simultaneously providing the funds for the same. It would provide the incentives for movement down development paths compatible with existing natural resources, while refusing to fund some countries' campaigns to mine their natural environments out of existence.

This would allow the trade between affected countries both to benefit the producers and consumers of goods but also to benefit those interested in the manner of development taking place in the producing state. For example, northern consumers could buy certified products (eg ivory) safe in the knowledge that they not only had acquired a beautiful natural product but also had helped to conserve the habitat of the African elephant. This is the sort of international regulation that can only be beneficial to the participants in trade: it is sorting out the lemons among competing producers of natural products, and this allows the interested green consumer to support the trade for reasons associated with the production process as well as the product itself.

It is down this path that CITES has to evolve if it is to become a workable system of international regulation. Over a period of twenty years it has achieved quite a lot in the development of systems that discriminate between different producers and allot quotas to those which are the most effective investors in their wildlife. There have been abject failures as well, but the important message here is that future international regulation must learn from both the failures and the successes of this regime. The remainder of this chapter will trace the path of the evolution of CITES into a form of constructive trade control mechanism.

THE ORIGINAL CONVENTION AS A TRADE REGULATION MECHANISM

As originally drafted, the CITES convention provided little in the way of a constructive trade control mechanism. The history of the CITES convention has witnessed many species progress from Appendix II to Appendix I, as potentially unsustainable trade levels raise concerns about the viability of the species. This progression occurred in the well-publicized case of the African elephant, for which a 12-year listing on Appendix II ended in 1989 with its uplisting to Appendix I by the Conference of the Parties.

Such a progression from 'potentially threatened' to 'endangered' is predictable, given the structure of the CITES convention. This is because an Appendix II listing gives little in the way of a wildlife trade control framework. An Appendix II listing leaves each range state operating independently, with no international assistance to perform the additional tasks that are required of the Parties or producer coordination to provide the incentives to conservation. Therefore, an Appendix II listing provides only additional tasks, and no real incentive framework, for the control of the trade in listed species. What an Appendix II listing certainly does accomplish is to publicize the potential rarity of the listed species. For some species this might actually result in an increase in consumer demand (for example, with regard to the exotic pet trade). Therefore, the consumer-side impact of an Appendix II listing is uncertain, and it could range quite widely.

Where Appendix II listing does encourage speculative purchasing, the combination of additional demand pressures with negligible control structures is a threatening one for Appendix II listed species. Since most wildlife species exist in unmanaged circumstances, it is usually difficult to handle the existing, let alone any increased, pressures. Thus, the progression of species from a listing on Appendix II to endangered status is not unforeseeable. For some species, it is entirely predictable. An Appendix I listing promises much more in the way of international cooperation. However, the efforts are not put to any constructive effect.

That is, once the regulated species completes the progression from virtually uncontrolled Appendix II species to endangered Appendix I species, the international community then launches into concerted action to ban the trade.

Of course, this will address the problem of possible immediate extinction from over-exploitation that might arise during an Appendix II listing; however, it does nothing to provide resources for the management of the species or the conservation of its habitat in order to avoid extinction in the medium term. For the endangered species of the developing countries, the withdrawal of value necessarily hastens the process of its elimination. They are under threat from both over-exploitation and habitat conversion. To address either of these forces, in anything other than short-term efforts, requires management and finance. In the worst cases, the situation has gone from bad to worse; with the uplisting to Appendix I, the traded species is protected from short-term extinction from over-exploitation, while simultaneously hastening the medium-term extinction of its habitat. Now, what is lost is not only the single species generating consumptive value, but also many of the related species and systems whose shared habitat could be subsidized by this value. Therefore, CITES, as drafted, provides for a peculiar sort of international regime of trade controls. For traded wildlife species it initially provides virtually nothing in the way of an international control structure, together with a global notice of potential rarity value (with the posting of an Appendix II listing), while following that with the withdrawal of developed world purchases of the wildlife product (via an Appendix I ban) if the population then comes under even greater pressures. The former, at best, provides no positive incentive framework; the latter provides no possible constructive use of wildlife value.

THE CONVENTION – ATTEMPTS AT INNOVATION

The Conference of the Parties to CITES has been taking steps toward a more constructive approach, with the attempted development of various sorts of constructive utilization systems. Although these are still in their formative stages, they represent the initial steps toward the recognition of the producer countries' perspective on the problem. At various times, important but not always effective, steps toward the construction of a rationalized international control structure have been taken.

The recognition of the need for constructive utilization

As early as 1979, the delegates from developing countries brought the anomaly of indirect extinction in lieu of direct over-exploitation to the attention of the Conference of the Parties. In San Jose, Costa Rica, they argued that there must be an economic benefit from the controlled use of species if they were to be able to justify protecting their habitats from development. These concerns gave rise to the first step towards the reform of CITES, with the adoption of Resolution Conf 3.15 at the New Delhi Conference of the Parties in 1981. This resolution provides for the downlisting of certain Appendix I populations for the purposes of sustainable resource management. The criteria which specify how Appendix I species may be utilized in order to procure compensation for their habitat are known as the Ranching Criteria, and each Conference of the Parties usually sees a large number of such proposals for review and possible acceptance. The first ranching proposal accepted involved the transfer of the Zimbabwean population of Nile crocodile to Appendix II in 1983 (Wijnstekers, 1988).

Ranching proposals tend to be focused on a particular state, or operation, and do not constitute mechanisms for the constructive control of the entire trade. In essence, they continue the ban in effect while allowing very limited, individual operations to recommence. While being of some utility, they do not constitute attempts at harnessing the value of an entire species for its own conservation.

In 1983, a species-based approach was first adopted with regard to the exploitation of the African leopard. Although listed on Appendix I, it was recognized in Resolution Conf 4.13 that specimens of the leopard could be killed 'to enhance the survival of the species'. With this, the Conference of the Parties approved an annual quota of 460 specimens, and allocated these between the range states. In 1985 this quota was then increased to 1,140 animals, and in 1987 to 1,830.

This approach to trade management was then generalized in 1985 with Resolution Conf 5.21, which provided for the systematic downlisting of populations where the countries of origin agree a quota system that is sufficiently safe so as not to endanger the species. Under this Resolution, five different species have been subject to quota systems: three African crocodiles, one Asian crocodile, and the Asian bonytongue, for which the Indonesians were allowed a quota of 1,250 specimens (the latter being a fish much admired by the Japanese as a wall hanging). None of these ranching systems went any further than the development of species-based quotas. In particular, no external control structure was ever implemented, this being left to the discretion of producer states. Thus, predictably, these quotas can be abused. For example, Indonesia is believed to have issued permits for about 140 per cent of its first year's quota of bonytongues (TRAFFIC, 1991).

At the seventh COP, in Resolution Conf 7.14, this scheme for developing quota systems was made time-limited, so that no quota system could continue beyond two COPs. The argument there was that CITES should encourage a movement away from general quota systems, and toward specific ranching regimes. This, however, is closely linked to the captive breeding movement. It is important to recall that it is only the farmers who benefit from farming what was formerly wildlife; conservation benefits accrue when harvest occurs in the wild (Luxmoore and Swanson, 1992).

The third avenue of early innovation under CITES was the creation of a management quota system (MQS) for the African elephant populations under Resolution Conf 5.12. This system was founded upon the ideas of management-based controls with consumer-based enforcement. Annual quotas were to be decided at the outset of each year, and producer states were then to issue permits not exceeding these quotas. Consumer states were then to disallow all imports unless accompanied by a MQS permit. This did not result in an effective control system for one very important reason. The management quota system provided no external checks on the discretion of the producer states. The determination of annual quotas and the issuance of MQS permits was within their unsupervised discretion. There were no externally-enforced incentives for sustainable use. This resulted in most states basing their annual management quotas of ivory on the expected confiscations from poachers. In addition, there were also no disincentives for cross-border exploitation, since consumer states were allowed to import ivory unquestioningly from any exporter issuing permits. Thus, Burundi, with one elephant, became the largest exporter of ivory in Africa under this control regime (Swanson, 1989).

The management quota system failed as a consequence of these clear inadequacies, resulting in a collapse of public confidence in the capacity for trade controls to work (Barbier *et al*, 1990). It is important to note that these control system failures are not costless. It is essential that an effective control system is developed and implemented before all consumer confidence is permanently lost in the potentially constructive capacity of wildlife trade. Nevertheless, despite the lack of enforceability, there is the germ of a good idea represented within these attempts. The use of national quotas that are linked in some way to the 'sustainable offtake' from existing stocks is one way of controlling the use of natural habitat. It is a mechanism that provides a clear and anticipated flow of funding to producers (by reference to the quota allocation) while it simultaneously provides the clear assurance of conservation to interested consumers (by basing quotas on stock levels). In this way, consumers who care are able to purchase from producers who conserve – effectively combining the two goods (production good and production process) into one.

The evolution of environmental conditionality

What continues to be necessary is a mechanism for the establishment and enforcement of such a sustainable quota system. A system that has been evolving in connection with, but not directly within CITES, addresses this need. This has occurred under the EU Regulation 3626/82, which effectively requires (under Article 5) an import permit for all Appendix II species. More importantly, for certain specified species (listed on Appendix C2 to the Regulation) there is an affirmative obligation on the exporting party to demonstrate that the export will not have a harmful effect on the population of the species in the country of origin. This Regulation has acted to move the costs of the monitoring and enforcement of sustainability from the producer to the consumer states, where it is more affordable. Producer states often find it desirable to mine their natural resources because it is far more expensive to manage the production process than it is to mine it (Swanson, 1994). This is why there is a widely noted positive relationship between development (income) and environmental management: the management of production processes requires resources that are not often available in the poorest of countries.

If developed countries care about effective resource management, and they are willing to provide the funds to support it, then there is no reason why they should not do so within the context of an international trade regime. The developed world's interest in the manner in which the flow of natural goods is provided (and its willingness to pay more for a flow that derives from a particular sort of production process) gives it the right to contract for a particular production process. This willingness to pay then allows for the imposition of the sort of conditionality that is required for the constructive use of the wildlife trade. The EU does in fact negotiate, on a country by country basis, the terms upon which it will remove that country's populations from Appendix C2. From this conditionality arises the possibility of state-by-state quotas created by the import permit obligation, and enforced by the EU customs inspectors.

In practice, the EU scheme operates by cancelling all trade in Appendix II species unless the exporter state meets its CITES Article IV obligation of demonstrating that the harvest is sustainable (or 'not detrimental to the wild population' in the language of CITES). When the exporter approaches the EU about the removal of the ban, the EU suggests a worldwide quota that would satisfy them as to the sustainability of the trade from that state. If the exporter agrees, then trade is resumed until the annual quota is satisfied.

The one objection to be made regarding the EU regime is that such absolute conditionality represents an imposition upon the national sovereignty of producer states. As mentioned earlier, the reason that a properly constructed trade mechanism should be unobjectionable is that there is no reason why trade need ever be disallowed under such a regime, only encouraged down particular pathways. The EU regime does not conform to this tenet, instead it disallows certain types of trade that should be within the sole discretion of another country. For example, is there any equitable basis for disallowing the Latin American states from trading in their tropical hardwoods, by countries that have cleared most of their hardwood forests long ago? If a country chooses to pursue a relatively hardwood-less development path (as many Northern countries have done long ago) should this not be its right?

In order to preserve the development choices of the producer, while simultaneously allowing consumers to have their say in the making of that choice, the trade regulation mechanism should act as an information system. That is, it should allow consumers to make informed choices about product/process combinations rather than requiring producers to deal in only one combination if they wish to trade. Then consumers are able to induce particular combinations of product/process by means of their willingness to pay higher prices for them. Producers are able to make their decisions concerning their stocks of wildlife by reference to the price differential applied to various management approaches.

The idea of a constructive trade regulation mechanism is to provide a mechanism by which consumers may discriminate in their purchases between different production processes. For example, a country that manages its elephants in a way that is certified as 'sustainable' could be allowed to sell its ivory through a mechanism such as an exchange, where consumers come to pay higher prices for the combined product and process (ie, ivory/sustainable management of elephants). On the other hand, a country that is rapidly removing its elephant populations should be allowed to sell its ivory on the world market, but only at the prices which unmanaged ivory is able to fetch. It is this discrimination between the two types of products that will allow every producer country to make its choice of development approach, while consumers are allowed to choose how much they wish to pay to influence those choices.

RESUMED TRADE IN IVORY – PART OF THE SOLUTION?

The various regimes that have been applied to the regulation of the trade in African elephant ivory were listed above: the Appendix II (permit-based) trade; the management quota system; and the Appendix I ban. The most recent ivory trade regime to evolve out of the CITES Conference of the Parties is a 'split-listing' regime, whereby several named southern African states have been allowed to resume Appendix II trading in ivory within a strictly controlled exchange. This exchange is to allow direct trade between these southern African states and their consumers in Asia (principally the Japanese).

The split-listing represents a piecemeal approach to the resolution of the problems within CITES. Instead, the Conference of the Parties should have adopted a more general resolution that created the capacity for country-based listings of all species, not just the African elephant. The crux of the issue is that the phenomenon of species endangerment is not a natural phenomenon, but a social one. It is a problem of resource management and a problem of a lack of resources to achieve first-best solutions in all situations. For this reason, the listings that should occur on CITES should not be the names of species but rather the names of nations that are not achieving 'sustainability' against some agreed criteria of management.

The difficult hurdle to cross in the implementation of a constructive trade regulation mechanism is the multilateral adoption of the criteria to be met for certification as a 'sustainable' producer of wildlife. The range of possibilities is endless. Producer states might reasonably argue that any stock level that they target produces a 'sustainable offtake', in that they demonstrate in the process that they are willing to apply the level of management that is required to achieve the designated stock level. On the other hand, consumer states might reasonably argue that only the maintenance of previously existing stock levels will produce a 'sustainable offtake', in that their preferences are based on the efficient management of existing stocks. The reconciliation of these very different points of view will require some difficult negotiations. The value of the development of an explicit set of 'sustainability criteria' is that it provides a clear incentive mechanism to induce states to invest in the management of their wildlife. The price differential achieved through managing in accordance with the criteria will induce states to manage in accord with those dictates.

In the medium-term there is little alternative to the development of such multilateral criteria. In the future, the various international environmental conventions will have to be brought into conformity with the GATT/WTO process, and this is what will be required. The WTO has expressly ruled that only multilateral regulation is allowed when production processes (rather than traded products) are the object of the regulation. When this principle is finally recognized in the case of CITES, it will be necessary for both producer and consumer states to agree a set of trade regulations. This does not bode well for the current consumer-led approach of enforced bans. The future of CITES is likely to be determined more by the need to develop regulations from the producer states' perspectives than the consumers', and this indicates the need for the recognition of the right to choose all sorts of development paths. In the future, CITES will have to become much more flexible, in order to allow producer states the choice of whether to trade within CITES (and receive whatever premium that confers), or not.

Therefore, resuming the trade in ivory on the basis of a split listing is only a small part of the solution to the CITES problem. What is required is that the convention attempt to generalize this idea into its fundamental workings. CITES must evolve into a certification mechanism, based on a set of multilaterally-agreed criteria for assessing the sustainability of national management regimes. Then only those states whose management systems are certified should be allowed to trade within the CITES exchange; all others should be allowed to trade outside of it.

CONCLUSION: THE DEVELOPMENT OF CITES

International regulation can only work if it is developed from the perspectives of all of the parties concerned. This might not seem to be a possibility in some cases. From the perspective of the average resident of a developed country, it would appear that the exploitation of wildlife is inconsistent with the conservation of wildlife. To a large extent, this perception is derived from a relationship with nature that has seen the extinction of a large part of the flora and fauna in the northern temperate zones. Other peoples in other parts of the world have not historically demonstrated the same relationship. For them, what we perceive as wildlife resources are in fact their primary resources. Many of them rely on species diversity and natural habitat in the same way that we rely on a few species such as cattle, sheep and domesticated fowl. For example, in many parts of rural Africa the vast majority of protein consumed still derives from wildlife sources. Given these differences in relationships, we have found ourselves in a polarized world. The Northerners have few remaining species; those in the tropics have the vast majority of the global total. Coincidentally, the countries with most species wealth also have the least material wealth.

This is the classical basis on which comparative advantage and gains to exchange are based. The difference lies in the fact that the developed world would be interested in paying countries in the developing world for the conservation of its stocks of natural resources as well as for the uniqueness of its flows. In that case it is not enough to have trade; it is also necessary to have a trade regulation mechanism which will inform consumers which flows derive from sustainably managed stocks. Then, the consumer is able to pursue both forms of trade at once: purchase of the relatively unique natural product together with the conservation of the region from which it came. A welfare-enhancing contract between North and South would enable both forms of exchange to occur.

CITES does not currently perform this role, but it is developing in that direction. The convention was initially drafted solely from a developed country's perspective. It provided for punitive measures against the trade when a species became endangered. It provided for no regulation, only monitoring, otherwise. The logic was simple, and Northern. Species endangerment equates with over-exploitation. Thus, all users of wildlife were punished, the sustainable with the unsustainable, when a species became endangered. This logic ultimately led to the absurd result at the 1989 COP when Zimbabwe, with a long and unquestioned history of sustainable elephant ivory utilization, was penalized along with the other ivory traders for the substantial decline of the African elephant population in a handful of states. The undiscriminating application of Appendix I resulted in the equal treatment of Zimbabwe (whose elephant population increased by 10,000 in the 1980s) with Tanzania, the Central African

Republic, Zambia and Sudan (whose joint elephant losses equalled about 500,000 during the same period) (Swanson and Pearce, 1989; Barbier *et al*, 1990).

The development of CITES requires two fundamental changes. The developed countries must learn to appreciate the perspective of the developing countries; wildlife utilization need not be inconsistent with wildlife conservation. Then, CITES must be reformed to discriminate between the constructive and the unconstructive use of the wildlife trade. That is, the objective is a constructive trade control mechanism that penalizes unsustainable utilization (relatively) by subsidizing the sustainable. The fundamental importance of all of the recent developments surveyed in this article (ie, ranching, quota and conditionality regimes) is that they represent a search for this middle ground between the Appendix I and Appendix II regimes. That is, they are the embodiment of the parties' recognition that the international regulation of wildlife trade may be turned to constructive effect in the developing countries. They also represent the first, halting steps toward the implementation of a contract that will recognize all of the Parties' trade objectives as legitimate. A trade mechanism should discriminate between differing trade flows, in order to inform consumers about what is happening in regard to the producer countries' stocks. In this way, the consumers are fully informed about the entire package of goods and services that their custom represents. This is the essence of effective and useful international trade regulation. It must be seen as regulation that will enable more complicated forms of exchange to arise, not simply as regulation to disallow the less desirable forms of trade.

REFERENCES

Barbier, E, Burgess, J, Swanson, T, and Pearce, D (1990) *Elephants, Economics and Ivory*, Earthscan, London

Diamond, J (1989) 'Overview of Recent Extinctions', in 'Population, Resources, and Environment in the Twenty-first Century', in Western, D and Pearl, M (eds) *Conservation for the Twenty-first Century*, Oxford University Press, Oxford

Luxmoore, R and Swanson, T (1992) 'Wildlife and Wildland Utilisation and Conservation', in Swanson, T and Barbier, E (eds) *Economics for the Wilds: Wildlands, Wildlife, Diversity and Development*, Earthscan, London

Lyster, S (1985) *International Wildlife Law*, Grotius, London

Marks, S (1984) *The Imperial Lion*, Westview, Boulder

Swanson, T and Pearce, D (1989) *The International Regulation of the Ivory Trade – The Ivory Exchange*, paper prepared for the International Union for the Conservation of Nature, Gland

Swanson, T (1989) 'Policy options for the regulation of the ivory trade', in Cobb, S
    (ed) *The Ivory Trade and the Future of the African Elephant*, Ivory Trade Review
    Group, Oxford
Swanson, T (1992) *The International Regulation of Extinction*, Macmillan, London
Swanson, T (1994) 'The Economics of Extinction Revisited and Revised', *Oxford
    Economic Papers*, Vol 24, p 85
TRAFFIC International (1991) 'Asian Bonytongue Exports from Indonesia', *TRAF-
    FIC Bulletin*, Vol 12, No 1/2
Wijnstekers, W (1988) *The Evolution of CITES*, Secretariat of the Convention on
    International Trade in Endangered Species, Lausanne
Wilson, E (1988) *Biodiversity*, National Academy Press, Washington
World Conservation Monitoring Centre (1992) *Global Biodiversity*, Chapman
    and Hall, London

Chapter 13

Decentralization, Tenure and Sustainable Use

Simon Metcalfe

LIVESTOCK VERSUS WILDLIFE

The prospects for the conservation of Africa's wildlife depend on the establishment of an efficient, equitable and sustainable system of community-based wildlife property rights. Such a system does not exist at present. The current distribution of ownership rights typically leads local people to favour livestock over wildlife.

Much of Africa (and two-thirds of southern Africa) is made up of dry savanna ecosystems. It is suitable as extensive rangeland for use by domestic livestock or wildlife. At present, most rural people prefer livestock management because domestic animals can be easily owned, used, marketed and are integral to the household production system. The state regards property rights over livestock as sacrosanct. Potentially, wildlife has economic and ecological advantages over lifestock, but in practice, it is not regarded as a valuable form of land-use because communities lack secure tenure rights over wildlife. Across the continent, local people have little formal standing in relation to rangeland or wildlife. In consequence, state, community and free-riding individuals (whether subsistence hunters or commercial poachers) typically contest the ownership of wildlife. In these circumstances local people will generally prefer livestock production to keeping land under wildlife. This creates a long-term threat to wildlife habitat, as there will be an incentive to convert that habitat to livestock use. This threat is likely to extend to protected areas.

The denial of proprietorship over wildlife to local communities also raises equity issues. The Maasai communities living on the Simanjiro Plains near Tarangire National Parks do not presently have the right to use the abundant wildlife on 'their' communal rangelands. They witness private companies licensed by the government marketing big-game hunts. Their permission is not sought, they receive no benefits, naturally feel angry toward the system that allows this and are hostile to the resource which uses their land yet gives nothing back. In the words of a Maasai elder, the government's wildlife policy is 'like a multistoried building without a ground floor'. So, the present system of tenure is inequitable and provides no incentive to local communities to conserve either wildlife or wildlife habitat.

THE ROLE OF THE STATE

The fact that local communities do not possess proprietorship over wildlife is a direct consequence of state policies. For most of this century the state has claimed formal proprietorship over wildlife. This has led to a 'command and control' style of conservation policy as the state has attempted to exert actual control over the resource it formally lays claim to. It has been very difficult for the state to succeed in this, not least because the local communities living closest to wildlife have had little reason to support the state's ownership claims. In consequence, illegal trade has sometimes threatened wildlife. This has been notably so in the case of elephants and rhinos. Local communities have not always been directly involved in this trade. In Tanzania and Zambia, during the 1980s, illegal commercial hunting was run by gangs rewarded by wealthy traders, often in collusion with corrupt government officials. Nevertheless, local communities often remained silent, in part because they had been warned off, but more importantly, because they had little stake in the ultimate fate of the wildlife. So, while commercial poaching has certainly been a threat to Africa's wildlife, it is not the underlying cause of species depletion. The fundamental reason lies with state policies that alienated wildlife from local landholders, while the state itself was unable to protect the resource.

BUILDING THE GROUND FLOOR

The conservation of African wildlife requires that the system of tenure be directly addressed. In particular, there are two elements that are necessary for any solution to the problems of habitat conversion and commercial poaching. First, there must be a clear devolution of ownership rights over wildlife to local communities. The unit of proprietorship should be the unit of production, management and benefit. Second, there must be an established market for wildlife products. The combination of clear resource entitlements and trade in wild species provides a positive incentive package to develop and conserve wildlife. Just as ownership of wildlife without trade would provide little incentive for conservation, so trade without well-defined ownership is insufficient to ensure sustainability. As long as wildlife is state property, local communities can not and will not invest in it. But, as communal property, wildlife can compete with or complement the use of communal rangelands for livestock.

Some states in southern and eastern Africa are devolving property rights in this way, although on occasion these attempts lack sincerity and commitment. State support is essential not only to achieve the transfer of property rights, but also to provide sanctions against those who would violate the new rights. The combination of state enforced negative sanctions and community-based positive incentives is the optimal solution to the problem of wildlife conservation. The shift from livestock production to wildlife management has implications for the distribution of benefits and burdens within local communities. Most livestock is owned by a minority of the community, who pay little for their access to forage and who have a strong vested interest in livestock production. A common property approach would be a threat to this rural power elite, be it a traditional or modern leadership. But, for the majority, such an approach would be attractive inasmuch as communal ownership would provide a redistributive mechanism from those who have stock to those who have the forage on which the stock depends.

Although livestock is differentially owned between households, local cattle 'barons' have customarily been accountable to the community. A fiscal arrangement, such as a community trust, could provide a formal approach to the issue. Users could be charged for a given period of access to the forage. In return, the community would have the funds with which to meet the costs of social security and a managerial control with which to insist on sustainable use. This model could be applied, not only to rangeland resources, but also to all natural resources where a defined user group wants access to communally owned resources. Tenure over common forage resources is at the heart of a sustainable multispecies approach to African savanna ecosystems. The worst case scenario would be a perpetuation of the blurred boundaries between state and community, democratic and traditional authorities, as well as between resource users and producers. In these circumstances only drought can assert control over stocking levels, humbling human management effort.

THE ROLE OF CITES

CITES assumes that international trade is a significant threat to species. When species are threatened it acts to restrict or halt trade in that species. For example, in 1989, when the poaching of elephants had reached unsustainable proportions (for some populations), CITES reacted by imposing a trade ban at the seventh COP in Lausanne. However, in responding in this way, CITES was treating the symptom rather than the underlying problem. It was the inadequate tenurial regimes of the range states that were the fundamental cause of the problem. Consequently, CITES policies of restricting or banning trade have limited effects in solving long-term problems. International trade controls provide no incentive for either local communities or range states to conserve wildlife habitat. Moreover, by banning trade in high-value species, CITES denies range states and local communities a vital source of revenue that might be devoted to conservation. The black rhino was placed on Appendix I in 1977, but, although trade was banned, the numbers continued to decline, because states had neither the will nor the incentive to devote the necessary resources to protecting them from poachers. Similarly, the elephant in southern Africa is the most prized species from a sustainable-use perspective, and the rent it can pay for its land use can help ensure the extensive range it needs is preserved. But, if local communities do not benefit from its conservation, then its range is threatened.

What CITES ought to do is to support both the devolution of tenure to local communities and a regulated trade in wildlife. At present, CITES provides no direct avenue for communities to express themselves except through their governments. It is assumed that states will represent the interests of local communities whether or not the communities have use rights. During the African elephant debate in Lausanne, community leaders who had been assisted to witness proceedings and give their views were astounded that the whole world could debate and determine the land use in their 'backyard'. Nor does CITES currently encourage governments to develop policies that devolve use rights to local landholders. When the Tanzanian and Kenyan governments argued for the ivory trade ban, there was no space in the CITES arena for communities in those countries, alienated from wildlife, to contend that they should have use and trade rights.

If trade is to be beneficial both to the conservation of wildlife and human development, then CITES has to address a total package of positive incentives (trade combined with secure tenurial arrangements) as well as protection (negative trade sanctions). Trade sanctions should be used to encourage the evolution of positive and efficient incentives aimed at conserving habitat and species. Trade sanctions on nations that are not accompanied by positive policy changes remove wildlife management as a land use option. More land for cattle is not what CITES should be about! Most southern African states are struggling to implement effective devolution of wildlife use rights and CITES should assist, not impede this. When local communities, range states and CITES work together the potential for positive policy and institutional synergy is much greater than when range and non-range states simply contest issues in the absence of local communities.

When a range state is practising good management, importing states should encourage trade; equally they should restrict trade with range states whose policies are patently not working. Trading states must share information to ensure policies are efficient, equitable and sustainable. A blunt instrument of punishing good and bad policies alike is unacceptable. The elephant trade ban, imposed in 1989, seriously undermined good management practices in southern Africa, while doing nothing to require that poor practices elsewhere were improved. No one should be fooled into thinking that the establishment of communal wildlife property rights is a simple exercise. But, in the end, it is more realistic and less problematic than attempting to conserve wildlife at national and international levels while neglecting the local landholders. A multistoried building of wildlife conservation must have a ground floor based on clear and unequivocal rights and responsibilities.

A Threat From Global Market Liberalization

One threat to the establishment of solutions based on the devolution of tenure comes from the global trend towards market liberalization. As liberalization advances, 'developing' countries have little choice but to enter the neo-liberal world of market economics and international trade. This drives poor countries, with large debts to service, toward export-led policies. For many parts of southern Africa this includes tourism, based on the region's unique and diverse wildlife. The danger is that this creates incentives to further alienate resources from communities in pursuit of joint ventures with foreign investors.

This has consequences for both equity and conservation. A state can use joint ventures to empower itself (politicians, bureaucrats and economic elites) at the expense of its rural people. If rural communities are alienated from their resource base their survival instincts could force an 'open access' resource scramble in Africa. The only way to counter the lure of joint ventures is to decentralize authority and responsibility over wildlife to local communities. Fortunately, in southern Africa there is some ground for optimism, as the policies in several countries reflect a desire to decentralize authority over natural resources and, as a consequence, we are witnessing a partial renaissance in relations between communities and their wild, open spaces. But, this is not the case in all of southern Africa.

LESSONS FROM SOUTHERN AFRICA

Two southern African nations present contrasting pictures. Mozambique appears vulnerable to the effects of liberalization, while South Africa seems to be more committed to devolution of tenure. In Mozambique there is no formal recognition of customary land rights for rural communities. The government can abrogate community rights of access to land and the natural resources in their neighbourhood at will. Although there was a community-based wildlife utilization experiment in Mozambique's Tete province, this required a special legislative diploma to permit the active participation of the community in a sport hunting enterprise. The general danger is that in a situation of state debt and structural adjustment, the government will mortgage the country's natural resources to foreign investors. This threatens indigenous local communities with no formal title to land, wildlife, forests, and coastal and marine resources. They may lose not only their present productive base but also their future land-use options. With the erosion of customary tenure, communities lose their negotiating power over resources desired by the state and international investors, and face the prospect of becoming merely a source of cheap labour. Moreover, even if local communities are legally alienated from their own resources, they are likely to try to assert their access to those resources. This forces the state into expensive and often futile protectionist approaches in order to enforce the exclusion of local people. At present, the continued presence of the tsetse fly in parts of Mozambique is doing more to conserve biodiversity (by preventing the conversion of wild habitat to livestock grazing ranges) than any national or international policy initiatives.

The present situation in South Africa is unique in post-independence Africa. The authorities have recognized traditional community land claims, going as far back as 1913. Consequently they have had to negotiate with communities who had been forcibly removed to make way for the establishment of protected areas. The most celebrated case involves the Makuleke community's claim to the northern part of the Kruger National Park. The national park authority has recognized the legitimacy of the claim and accepted a co-management arrangement by which benefits will flow to the community. In return, the community has accepted that the land remain within the overall area protected as wild land. The protected area status remains but the participation in management costs and benefits are both local and national. Communities do not just benefit from activities outside the park but from those inside as well.

Outside protected areas, community land in South Africa, as in most of southern Africa, is legally state land, but the South African government recognizes that, whatever its legal status, the underlying right remains with the community, whether traditionally or democratically defined. The Minister of Lands is the nominal owner of the land on behalf of the government but intends to transfer the land in an 'orderly and transparent manner' to local communities. Pending the finalization of land transfer, investors can obtain legal security of tenure through long-term leases that will be registered in their favour by the Minister of Land Affairs as the nominal owner of the land. Once the land is transferred to communities from the state, they will 'step into the shoes' of the minister, inheriting all his legal rights and obligations relating to the land. From the outset, communities will be intimately involved in the negotiation process to ensure they are satisfied with all the agreements. The underlying principle is that communal land rights are retained by the users (ie, the farmers) and not the institutions that represent them, traditional, democratic or statutory. Elsewhere in the region the situation is that land is held through the institutions first and the users second.

This example indicates a way through the present impasse around communal tenure. Rather than the state alienating communities in favour of partnership with the private sector, the state acts in partnership with communities first and works with them in negotiations with those who want private access to the resource. The intent is clear – the state recognizes communal tenure, facilitates its evolution and underwrites its interests. The state's land policy and its spatial development plans can be linked directly to the tenurial interests of local communities. Assuming that the state is acting in good faith, this appears to be a viable approach. Nevertheless, many of the problems of implementation remain to be solved.

CONCLUSIONS

Successful conservation requires the successful devolution of ownership rights over wildlife to local communities. This, in turn, requires the support of both individual national states and CITES. These two can combine to provide the combination of positive incentives and negative sanctions that can make the policy a success.

This would require a considerable change in CITES. CITES cannot continue to look at trade in isolation from sustainable use issues such as communal wildlife property rights. Trade without secure tenure rights is unlikely to be sustainable because the basis of efficient management and equitable distribution of costs and benefits would be absent. Tenure may not be a sufficient condition for sustainable use but it is a vitally necessary one. Unless equity within a generation is addressed, equity between generations (sustainability) cannot be achieved. There may also be tensions between local communities and national states. In particular in the context of market liberalization, local communities should be prepared to defend their newly devolved rights against national elites who may use their power to alienate communities from their natural resources in order to secure short-term gains for themselves. Local communities might even look to a reformed CITES for support in the face of such conflicts. In any case, CITES must either address these issues or accept that it is as much a part of the problem as a part of the solution.

Chapter 14

Global Regulation and Communal Management

Barnabas Dickson

INTRODUCTION

During the colonial period in Africa a particular approach to wildlife conservation emerged. It assumed that the main threat to wildlife came from the human exploitation of wildlife and that the appropriate policy response lay in the creation of protected areas and wide-ranging restrictions on hunting. CITES, which was first signed in 1973, accepted some of the assumptions of the colonial approach. It treated the international commercial trade in wildlife as the chief threat to many species and it imposed further restrictions on that trade on top of the existing domestic restrictions. This approach to wildlife conservation has been coming under strain in recent years. The enforcement of restrictions has become increasingly difficult and some protected areas have become the stamping grounds of poachers. It has also become apparent that, for many species, the major threat comes, not from trade, but from the loss of habitat. This has led some to propose a new way of conserving wildlife, based on the notion of sustainable use. The central idea is that unless wild species can provide benefits to humans they will not be conserved in the long run.

The debate between the opponents and proponents of use has often been fierce. But, within CITES, the supporters of sustainable use do seem to be gaining ground. At the most recent COP, in Harare in 1997, a decision was made to partially revoke the eight-year ban on the international trade in ivory, and to allow Botswana, Namibia and Zimbabwe to sell some of their stockpiles. If this decision signals a lasting shift, then it could also herald the start of a new debate. This will not be about the pros and cons of use, but about the way in which use is to be made sustainable. The new debate is a potentially complex one, as those who favour sustainable use come from a wide range of positions. But, there are two versions of the pro-use argument that are particularly prominent and this chapter focuses on these.

On the one hand, there are those who emphasize the necessity for the global regulation of the international trade in wildlife. The purpose of the regulation should be to ensure that all such trade is sustainable. On the other hand, there are those who stress the communal management of wildlife as a common property resource, as the key to the conservation of many species in the South. The latter group focuses its attention on the importance of granting proprietorship over wildlife to local communities, and points to other benefits, besides the conservation gains, that this type of policy can realize. The CAMPFIRE programme in Zimbabwe is frequently cited as a successful example of communal management.

The two positions have different implications for the future evolution of CITES. The global regulation camp require a system of enforcement for their regulations and suggest that a suitably reformed CITES could undertake a key role in this. The advocates of communal management contend that most of the responsibilities for enforcement should rest with local communities, and they see a less significant role for CITES.

As long as the debate with the opponents of the trade in wildlife was at the centre of attention, there was an incentive for the proponents of sustainable use to gloss over their differences. But, if the Harare decision signifies a widening consensus on the value of trade as a conservation tool, then the differences among those who support use are likely to assume a greater prominence. This chapter is intended as a contribution to the comparative assessment of the global regulation and communal management positions. The chapter begins by looking at the colonial approach to conservation, since this has been such a major influence in determining policies and shaping perceptions. It then turns to CITES and the course of debates within it, moving to a discussion of the two pro-use positions. The final section assesses the relative merits of each.

THE EVOLUTION OF THE COLONIAL APPROACH TO CONSERVATION

In recent years there has been an increased interest in environmental history. Some of this work has emphasized the connections between

European imperialism and the development of conservation policy (eg Grove, 1995). This is particularly true of wildlife conservation, where John MacKenzie has produced an important study (MacKenzie, 1988). The research of MacKenzie and others is valuable not only because it serves as a useful reminder that current debates about endangered wildlife are not as new as they sometimes appear. It also throws light on the debates themselves, exposing inherited assumptions and showing how past policies have circumscribed subsequent choices. MacKenzie contends that the history of wildlife conservation in the British Empire is intimately bound up with the history of hunting. He argues that hunting went through several stages, but particularly significant was the emergence of the idea that hunting was to be pursued as a sport, rather than for more pragmatic reasons. By the beginning of the First World War hunting for sport had become the norm in many British colonies. As a sport, it involved a quasi-ethical orientation towards both wildlife and the practice of hunting itself. Some species were regarded as worthy quarry and certain ways of hunting came to be viewed as appropriate and fitting, while others were dismissed as unsporting and cruel.

Closely linked to the development of sport hunting in the colonies was a growing belief in the need for conservation. The earlier periods of European hunting had resulted in a considerable reduction in animal numbers. It was felt that something needed to be done if sufficient animals were to be available for hunting – even as a sport – to continue. Importantly, conservation was conceived as a task for empire. This is nicely illustrated by a remark of Lord Curzon's in 1906. A leading figure in the Society for the Preservation of the Fauna of the Empire, he expressed the belief that 'We are the owners of the greatest Empire in the Universe... trustees for posterity of the natural contents of that Empire' including its 'rare and interesting animal life' (MacKenzie, 1988, p 213). This combination of ethical purpose with a sense of imperial responsibility exemplified the conservationist rhetoric of the time.

In practical terms, this new conservationist mood led to a plethora of regulations in the colonies, successively reducing the categories of those who were entitled to hunt. These regulations were supplemented with the creation of game reserves. In these reserves, often established at around the turn of the century, hunting, except of those animals characterized as vermin, was prohibited. The thinking was that these reserves could ensure a supply of quarry for European hunters. This marked a crucial stage in the evolution of conservationist thinking. It saw the emergence in Africa of the notion that the conservation of wildlife was incompatible with the presence of human communities. Separation was necessary. The conservation measures also had an important social dimension. They served as a way of reinforcing social differentiation within the settler communities. Hunting, as a sport, was becoming the preserve of the social elite and the conservation of wildlife was at the same time the conservation of a particular social order. Hunting contributed in other ways too, with the revenues from tourist hunters in east Africa contributing significantly to the sometimes hard-pressed colonies (MacKenzie, 1988, p 149).

Whatever the role of hunting and conservation within the settler communities, its impact on the indigenous populations was more drastic. Prior to colonialism, hunting had played an important part in the lives of most Africans, but the colonial authorities banned almost all hunting by Africans. When reserves were created the African populations were removed, often by force. Thus, conservation involved a rearrangement of property relations. Game reserves became state land and game became state property. The imperial state's assumption of a quasi-moral responsibility for conservation was implemented through the state appropriation of wildlife. This was all of a piece with the racial separation that was gradually being imposed. Africans were displaced from much of the land they had formerly occupied. Broadly speaking, that land became either the private property of European farmers, or it entered state ownership and was used for conservationist purposes. The lack of legitimacy that these measures had among indigenous people, coupled with the fact that they had little incentive to conserve wildlife resources over which they no longer had any rights, placed Africans in a position of potential conflict with conservationist goals. This was to have long-term implications.

The early conservationist measures were linked to satisfying the needs of European sport hunters. In time, however, a change in the rationale took place. The view that wildlife should be conserved, not for shooting, but for viewing, won increasing acceptance. This change was closely linked to the establishment of national parks in many of the colonial states. The Kruger National Park in South Africa was established in 1926; and in 1933 the Agreement for the Protection of the Fauna and Flora of Africa was signed in London. This agreement made an explicit call for the establishment of national parks and the call was widely heeded both before and after the Second World War.

In some ways, the shift towards conserving wildlife so that it could be viewed, rather than hunted, marked an important change in the colonial approach to wildlife. In other respects the continuities were strong. This was particularly so for Africans. As with reserves, the establishment of national parks often involved their forced removal and further prescriptions on their hunting and grazing activities (see Ranger, 1989; Bonner, 1993). In both cases, their needs and interests were of little or no account.

Conservation policy also continued to play an important social and economic role among the European settlers. But, instead of reinforcing social distinctions, part of the function of the national parks (as the term itself indicates) was to help with the elaboration and strengthening of a new sense of national identity that included all the settlers. Carruthers, for example, argues that the establishment of the Kruger National Park (named after Paul Kruger, a former President of the South African Republic) played a significant role in uniting Afrikaans and English speakers, and in cementing cross-class alliances among the whites, alliances that had, if anything, been endangered during the earlier period when hunting was a sport pursued by the elite (Carruthers, 1989). Similarly, Ranger has pointed out that one powerful impulse motivating the creation of a national park in the Matopos (in what was then Southern Rhodesia) was the fact that Cecil Rhodes' grave is located there (Ranger, 1989). The revenue-generating potential of the new conservation policy was also important. Conceived at the start of the new era of motorized tourism, Carruthers argues that the South African proponents of the Kruger park had been influenced by the economic success of national parks in the US (Carruthers, 1989). Their role as a source of revenue was to be important in the post-colonial era.

In showing how colonial conservation policy emerged from the practice of hunting, MacKenzie and other historians have demonstrated that that policy was not just about wildlife. It was shaped by economic interests, beliefs about race and social order and views on the appropriate way to relate to the natural world. As MacKenzie puts it, hunting and conservation can be seen as 'a complex network of economic, social, racial and cultural relationships' (MacKenzie, 1988, p 2). The possibility that there may be a wide range of factors motivating conservation policy is often lost from view in contemporary policy discussions. The colonial approach to conservation provided the starting point for all future conservation efforts in the post-colonial societies. Subsequent policy often involved important continuities with the colonial era and even when later conservationists advocated the rejection of that legacy, it was the colonial approach against which they were reacting.

CITES

As African countries began to gain independence, there were fears among European conservationists that the existing system of conservation would collapse. Noel Simon, a settler in Kenya and writer on wildlife, expressed the view in 1962 that 'the notion of conserving the creatures of the wild to ensure their continuance into the future is alien to the African' (Bonner, 1993, p 64). Even more significant were the views of Max Nicholson, a founder of both IUCN and, later, WWF. He helped to draft the Arusha Manifesto that President Nyerere of Tanzania delivered in 1961. The manifesto was partly the outcome of an attempt by Western conservationists to get the new African leaders to commit themselves to the conservationists' agenda. Nicholson explained their motivation in the following way: 'We felt that under the new African governments, all prospect of conservation of nature would be ended' (Bonner, 1993, p 64).

It was in this climate that the IUCN initiated discussions during the 1960s about establishing a treaty to regulate the international trade in wildlife. This led, eventually, to the signing of the Convention on International Trade in Endangered Species in 1973, in Washington DC. The treaty instituted a system of regulations for the international trade in wildlife, including bans on commercial trade for the most endangered species. In effect, it imposed a further set of controls on those already existing at the national level. It represented the view that, with doubts about the commitment of the newly independent states to conservation, the existing system – with its protected areas, state ownership of wildlife and bans on most forms of hunting and trade – needed strengthening from above. It was an attempt to shore up the colonial approach to conservation in a post-colonial world. An international treaty was the ideal vehicle for this since, while it required the consent of the former colonies, there was an expectation that, with their greater resources and experience, the countries of the North would be able to shape the direction and operation of the treaty. This expectation was borne out in the early years of CITES.

Associated with these changed circumstances was a change in the way in which the responsibility for conservation was conceived. It was no longer to be understood as an environmentalist version of the 'white man's burden', of the sort articulated by Lord Curzon. Instead, it lay in the ostensibly more progressive idea of conservation for the sake of all humankind. The first clause of the preamble of CITES stipulates that states have a responsibility to protect wild fauna and flora 'for this and the generations to come'. Subsequent supporters of CITES have not infrequently invoked the common heritage of human-kind principle in order to justify states in the North devising regula-tions governing the use of species that are located in the states of the South (Favre, 1993, pp 891–3).

The system of controls established by CITES centres around two appendices on which species can be listed. Appendix I is for the most seriously endangered species and involves a ban on international commercial trade in the species in question. Appendix II is for less threatened species and, while trade is allowed in these species, the assumption is that it is potentially dangerous. There are provisions that are designed to ensure that such trade is not detrimental to the survival of the species. However, the decision on whether trade is detrimental is left to the exporting countries, with no mechanism for external scrutiny. As a result, these controls have often not worked and many species have been moved from Appendix II to Appendix I. Moreover, many Appendix I species have themselves continued to decline. Some have responded to these difficulties by claiming that what is needed is more effective enforcement, with stronger action against illegal traders and the states who tolerate them. This presents the problem of conservation as a problem of criminality. Indeed, the history of CITES has been punctuated with calls for more effective enforcement, sometimes coupled with stories about the work of in-ternational criminal gangs in the wildlife trade. Others have argued that what is needed is a re-think of the more fundamental assump-tions of CITES, in particular, the grounding assumption that trade is always likely to be a threat.

Some of the calls for a more positive approach to the trade in wildlife have undoubtedly come from trade interests, who have little concern with long-term conservation. Others have come from con-servationists who genuinely maintain that trade can be beneficial. This type of argument has had some effect on CITES. While the original articles of the treaty make little provision for an internation-ally controlled trade in endangered species (Appendix I rules out such trade and Appendix II imposes largely ineffective controls) the Parties have been able to move in that direction by annotating list-ing decisions in various ways, to include quotas for trade and split listings for different populations. These moves have faced vigorous opposition from some quarters and the balance has swung back and forth, but since those who favour sustainable use have been having increasing success, it is worth looking at their arguments in more detail.

GLOBAL REGULATION

There are different views on how best to ensure that the use of wildlife can become a positive tool for conservation. The remainder of this chapter will focus on the two most popular of these views. Both share the belief that unless human society can benefit, in some way or other, from the use of wildlife, then that wildlife will be under threat. They therefore place the use of wildlife at the centre of their conservation strategies, but the two views differ in their proposals for ensuring that such use is sustainable.

The global regulation approach, which is the subject of this section, places a large emphasis on a reformed system for regulating the international trade in wildlife. The basic idea is that the global community, under the leadership of the states of the North, should regulate the wildlife trade in such a way as to encourage trade in wildlife where it is sustainable, and to discourage or penalize it where it is not. It is held that they are in a position to do so because much of the ultimate consumption of wildlife products takes place in the North. These states can therefore encourage sustainable use by such means as buying only from sustainable producers, or offering them a premium above that paid to unsustainable producers. Since much of the world's remaining biodiversity is located in the South, the trade regulations will be directed mainly at Southern states. It is a feature of this approach that it assumes that the key agents for ensuring the use of wildlife is sustainable are states, individually and collectively.

This positive view of the international trade in wildlife marks a decisive shift from the colonial, protectionist approach to conservation that influenced the formation of CITES. Instead of trying to restrict trade on the assumption that it is always potentially threatening, the global regulation approach aims to supply the producer states in the South with incentives to make trade sustainable. The purpose is to exploit their self-interest for conservationist ends. In the early part of his book, *The International Regulation of Extinction*, Timothy Swanson develops a theoretical model that provides a defence of the global regulation approach (Swanson, 1994). The model explains why states in the North and the South will be disposed to accept the appropriate type of global regulation of the wildlife trade as part of the solution to the problem of declining biodiversity.

Swanson holds that the fate of wildlife is now determined by humans. More particularly, it is determined by 'a societal-level determination' of the investment-worthiness of wild species (Swanson, 1994, p 11). A society will invest in a species if it offers a competitive return.

As he puts it, 'If a society views a species as sufficiently investment-worthy, then stocks of the species will be maintained (through adequate levels of investment in it and its ancillary resources). If the species is not viewed as relatively investment-worthy, then it will suffer disinvestment... and ultimately extinction' (Swanson, 1994, p 11). He holds that there are several different routes through which a species might become extinct, but the end result is the same. Swanson assumes that the state acts straightforwardly as a vehicle for implementing the societal-level determinations of the investment-worthiness of species. That is, the actions of the state are assumed to issue from a desire to promote consequences that are socially optimal. As he expresses it, 'societal objectives will be inherent in state decision-making' (Swanson, 1994, p 11).

Since biodiversity is now concentrated in the South, its fate turns on whether the states of the South view species as investment-worthy. In Swanson's view wild species currently do not offer an attractive investment opportunity. It will often be in the interests of these states to convert the remaining areas of wildlife to other, more productive, uses. From their point of view this is the economically optimal course of action. However, from the global point of view, this will not be the optimal decision, for such diversity has great value (including both information and insurance values) and that value increases as the amount of remaining diversity is reduced. The problem is that the states of the South are not appropriating this value and the costs of its destruction are not fully internalized. They therefore have little incentive not to convert. Conceived in this way the threat to biodiversity is a problem of global externalities that arises because the world is divided into independent states. Those states make decisions that are optimal from their point of view but which result in globally sub-optimal outcomes. He writes, 'these "international externalities" are the globally applicable and systematic reason for diversity decline' (Swanson, 1994, p 11).

Once the problem has been conceptualized as a problem of externalities it follows that the key to the solution lies in correcting these externalities. Thus it is that Swanson advocates the establishment of institutional frameworks that will ensure a transfer of the values of diversity from North to South. If this can be achieved then it will alter the incentives faced by the states of the South so that their optimal course of action will be to invest in their diverse resources. This will lessen the threat to those resources and so bring about something closer to the globally optimal result. Swanson's general model of the source of the threats to biodiversity appears to explain why the type of global regulation of the wildlife trade that he advocates is a viable way of dealing with the problem, since it explains why the states of the North will be motivated to establish such a system of global regulation. This regulation will bring about the conservation of biodiversity, which is optimal from their point of view. It also explains why the states of the South will then invest in species. The system of global regulation will make it optimal for them to do so.

However, questions can be raised about Swanson's model. Like many models it relies on some simplifying assumptions, and if these assumptions diverge too far from reality then the model will be of little practical value. Central to Swanson's model is the assumption that individual states pursue policies, including policies towards wildlife, that are socially optimal in the economists' sense. It is this assumption that allows him to analyse the problem as one of externalities. He admits that 'this implies a fairly innocuous vision of the relationship between individuals, society and the state' (Swanson, 1994, p 11). He goes on to suggest, however, that this simplifying assumption is not of any great import since he is only concerned to show that if states did pursue socially optimal policies this would not, because of externalities, result in a globally optimal solution (Swanson, 1994, p 11). However, this would seem to understate the limit of his ambitions, for he does not restrict himself to this hypothetical claim. He relies on the assumption about optimality both in his explanation of why species extinction has occurred and in the general form of his policy prescriptions. So, if that assumption is mistaken, then the explanation and the prescription will be called into question.

The assumption that individuals always pursue their economic self-interest has often been queried. The parallel assumption that states pursue socially optimal policies seems even more doubtful. Even the brief discussion earlier of colonial conservation policies showed that state policy toward wildlife was shaped by a host of different factors, including beliefs about race, social order, imperial responsibility and economic interest. It seems likely that the conservation policy of modern states, both North and South, may be shaped by a similarly wide range of considerations. Moreover, even if states were disposed to pursue wildlife policies that were socially optimal, it is not clear that they possess the necessary information to allow them to do so. The economic values of diversity are very difficult to establish.

If state policies towards wildlife are not motivated by a desire to promote socially optimal outcomes then Swanson's model is no longer applicable to the real world. It does not provide a reason to think that the states of the North will support this type of global regulation, or that such regulation will cause the states of the South to invest in wildlife. However, it is important to see that this criticism of Swanson's model does not imply that the global regulation approach to conservation is itself flawed. All it shows is that states will not have that particular motive for promoting the global regulation solution, but they might have other reasons for doing so. Indeed, Swanson himself makes some suggestions about other reasons when he considers the global regulation solution in detail. Whether or not there are such reasons will be considered in the final section of this chapter.

In the meantime, it can be noted that the global regulation approach to conservation implies a large role for a suitably reformed CITES. For CITES could become the global institution that would regulate trade in the appropriate way. This is certainly how Swanson sees it. He argues that CITES has actually moved, falteringly, in the direction of establishing the appropriate type of trade regime and that this is a trend that can be fostered (Swanson, 1994, Chapter 8). A related point is that while the pro-use element in the global regulation view signals a break with the colonial, protectionist approach to conservation, the fact that ultimate responsibility for conservation remains in the hands of the states of the North marks a continuity with the colonial legacy. The second pro-use view is different in this respect.

COMMUNAL MANAGEMENT

The communal management approach to making the use of wildlife sustainable shares with the global regulation view the assumption that the key to successful conservation lies in ensuring that wild species benefit human society. But, it provides a very different account of how this is to be done. It relies, in the first instance, not on the global regulation of the international wildlife trade, but on giving local communities in producer countries proprietorship over wildlife. The reassignment of proprietorship from the state to local rural communities is the linchpin of this approach. Murphree refers to the 'institutional centrality of proprietorship' (Murphree, 1994). The idea is that once local communities have proprietorship over wildlife they will be able to benefit from those wild species. They will be able to trade in wildlife products, sell licences to safari hunters, profit from wildlife-based tourist ventures and engage in subsistence hunting. This will provide them with a strong incentive to conserve the resource for the future.

The emphasis is on *communal* management because this is in keeping with traditional forms of social organization, existing socio-political realities and the fugitive nature of the resource.

It is an important aspect of the communal management approach that where it is successful it improves the welfare of poor rural communities and empowers them. The aim of those who advocate the communal management of wildlife is both to institute more effective means of conservation and to do more to meet the aspirations of local communities. The weight placed on this redistributive goal is one feature that distinguishes this from the global regulation approach. Another difference is that while global regulation placed the primary emphasis on states (and the global community of states) as the appropriate agency for ensuring sustainability, the communal management approach sees rural communities as the key agents.

There are two main sources for the communal management view. The first lies in an explicit critique of the colonial approach to conservation. As was seen, that approach involved the state assuming ownership of wildlife and instigating widespread restrictions on the use of wildlife. The proponents of communal management argue both that this did not work as conservation and that it was unjust. It did not work because the rural people living closest to wildlife had little incentive to conserve wildlife. Since they had no legal claim on that wildlife they foresaw little long-term gain from it. On the contrary, it was often a threat to their livelihoods (when wild animals destroyed their crops) and sometimes to their lives. They had no reason not to acquiesce in poaching and positive reason to engage in the practice themselves. In these circumstances, it should not have been surprising that state attempts to protect wildlife often ended in failure. The colonial approach was condemned as unjust because the colonial authorities had deprived indigenous people of a valuable resource that, prior to colonization, they had regarded as their own. In addition, the state typically sought to protect wildlife under its nominal ownership by the use of extremely harsh methods, including the extra-judicial execution of suspected poachers.

The advocates of communal management hold that both these defects can be remedied by restoring ownership rights over wildlife to local communities. This places control over the fate of wildlife, at least in the first instance, in the hands of rural communities. Thus, the communal management approach does represent a more radical break from the colonial approach than the global regulation view, which assumes that the guarantors of the safety of wildlife are the states of the North.

Nevertheless, the communal management approach is not free of all trace of the colonial era. In Zimbabwe, for example, the CAMP-FIRE programme (which is often cited as an example of a successful communal management scheme) has relied on income from safari hunting for a large part of its revenues. The safari hunters are mainly from Europe and North America and they are prepared to pay large sums to hunt game in Africa. Safari hunting, of course, is an activity emblematic of the colonial era, and this fact is often seized on by critics of CAMPFIRE. However, the reliance on safari hunting might be better seen as a reflection of two facts. The first is the continued global inequality that makes safari hunting an attractive proposition for poor communities seeking to make money from wildlife. The second is the enduring influence of the colonial approach to conservation. It has had a profound influence on the attitudes, practices and institutions connected with wildlife, and the legacy is difficult to completely cast off.

The second source of the communal management approach is recent theoretical work on common property (eg Ostrom, 1990). The starting point of this work is typically a critique of Garret Hardin's *Tragedy of the Commons* (Hardin, 1968). Hardin presents this tragedy as if it was a problem for any resource managed as common property. The implication is that the solution lies either in the privatization of the commons through the creation of private property rights, or in the establishment of some external authority to regulate the use of the resource. However, common property theorists argue that Hardin has misrepresented the issue and that the problem he discusses is really one of a *de facto* open access resource. They would claim that the difficulties that open access gives rise to do nothing to cast doubt on the validity of common property solutions. On the contrary, they would argue that common property regimes are often more successful in dealing with problems of open access than solutions based either on private property or external authority. Indeed, it is often the failure of external authorities to enforce their nominal control of an environmental resource that gives rise to open access problems in the first instance. This has certainly been true of wildlife resources during the colonial and post-colonial eras. This last point indicates how closely this theoretical work on common property has meshed with the first source of the communal management approach: the critique of colonial conservation. Some authors draw explicitly on the two sources (eg Bromley, 1991).

Consistent with its more radical – if still incomplete – break with the colonial legacy, the communal management approach implies a rather different role for CITES. It assumes that rural communities will guarantee the future of wildlife. CITES will not be the primary vehicle for saving wild species from extinction. Its function will be to provide an appropriate regulatory framework that will support local communities in their efforts to promote the sustainable use of wildlife.

ASSESSMENT

If the proponents of the sustainable use of wildlife are gaining the ascendancy over those who are opposed to all use of wildlife, the differences between the global regulation and communal management approaches to sustainable use are likely to assume a greater importance. Neither approach is free of weaknesses.

Three queries about the global regulation approach can be raised. The first concerns the motivation of the consumer states of the North. Swanson outlined a theoretical model that explains why those states will be disposed to support a regulatory system that rewards sustainable use. The model relies on the assumption that states pursue policies that are, from their point of view, socially optimal. But this assumption is doubtful, and it is therefore crucial to know if those states have any other reason to support such a system. Their historical record is not impressive. A considerable proportion of their biodiversity has been lost and the early stages of their colonization of other parts of the world were often marked by the widespread destruction of wildlife. Of course, it can be argued that their attitudes have now changed. In *The International Regulation of Extinction* Swanson mentions two reasons they have for supporting global regulation. The first is that as the ultimate consumers of wildlife products they have an interest in its long-term conservation, for they will want to ensure a continuity of supply (Swanson, 1994, p 213). However, even if it is granted that, for example, those involved in the working and sale of ivory in Japan do have an interest in assuring the long-term supply of ivory, it does not follow that the Japanese state will be sensitive to their interests, or that it will not have other reasons for taking a different stance. The second reason Swanson cites is that many individual consumers in countries of the North are committed to the conservation of wild species and that therefore they will support sustainable use (Swanson, 1994, pp 224–5). But, as Swanson himself seems to acknowledge, in so far as consumers are committed to conservation, they are also often opposed to the consumption of wildlife products. The belief that such consumption is likely to endanger rather than protect species, coupled with the contention that it is, in any case, morally objectionable, has a deep hold. So, support for conservation may not issue in support for sustainable use. Once one rejects the simplifying assumption that states are motivated by a desire to promote socially optimal results and one examines the actual factors that can influence state policy in the North it is not obvious that one should expect consistent support for sustainable use.

A second weakness of the global regulation approach is that it makes enormous information demands on the states of the North. If the system of regulation is to work then these states must be able to make accurate determinations, in a wide range of cases, of whether trade in wildlife products from the producer states is sustainable. This is a notoriously difficult thing to do and uncertainty is the rule rather than the exception. This makes it difficult to implement global regulation in practice.

A final consideration concerns the justification of the states of the North imposing environmental conditionality on states of the South. If it was the case that the system of global regulation could reasonably be expected to realize a globally optimal result, then there would be one defence available to its proponents, although such a defence would only be as strong as the claim that what is economically optimal is also ethically desirable. But, once the claim about promoting the globally optimal result is discarded, the states of the North may have to appeal to notions such as the common heritage of humankind in order to justify their imposition of trade regulations. Such appeals are often treated with some scepticism in the former colonies of European powers.

None of the above points show that the global regulation approach to conservation is fatally flawed. Nevertheless, the communal management approach does appear to be able to sidestep many of these difficulties. Since it grants proprietorship over wildlife to local communities it does not require, at least in the first instance, that the states of the North are motivated to promote sustainable use. For the local communities will themselves have sufficient incentive to ensure that use is sustainable. Their livelihoods will depend on it. These communities, with their local knowledge and long experience, will also find it much easier to determine when their use is sustainable. Moreover, the communal management approach does seem more desirable on grounds of social justice and the charge of neo-colonialism is less easily laid at its door.

Despite the advantages of attempting to promote sustainable use by creating appropriate institutions at the local level rather than by imposing a system of global regulation, however, it does not follow that the wider context is irrelevant to the success of communal management. First, since that approach rests on a reassignment of property rights from the state to local communities, it is important that the state pursues this wholeheartedly. A failure to do so can undermine the success of the enterprise. In practice, this has sometimes been a real stumbling block, with states in the South not eager to cede too much control to local communities.

Second, even if local communities do gain full proprietorship over wildlife, there is no guarantee (as Swanson might point out) that it will be in their interests to conserve the wildlife. They might do better to mine the resource and invest the returns elsewhere. It will depend on factors such as the price they receive for wildlife products and the return on alternative land uses. These factors will be well outside the control of local communities.

Third, there are a range of socio-economic processes that can simply overwhelm communal management schemes. For example, in many countries of the South the scale and speed of social change often results in large population movements. A population influx into an area can make the sustainable management of a natural resource an impossible task. A rather different sort of possibility is that, in situations where wildlife does indeed constitute a valuable resource, communal management schemes will be promoted for public relations purposes, while larger economic interests exploit the wildlife for their own benefit.

Again, none of these weaknesses shows that communal management cannot work, and the attractions of this approach are considerable. The historical evidence suggests that no conservation policy will work unless the structure of incentives among local communities that live closest to wildlife favours conservation. Moreover, when the effect of past conservation policies on indigenous people is properly recognized, the record is a shameful one. To the extent that those policies were motivated by a moral concern with protecting wildlife for future generations, it was a peculiarly blinkered type of morality. It is a strength of the communal management approach that it addresses this concern. Finally, some of the weaknesses in the communal management approach could be rectified by combining it with an appropriate system of global regulation. Such a system could go some way to ensuring that there is a market for the products of sustainable management.

It is a merit of both the global regulation and communal management approaches to conservation that they acknowledge the importance of placing conservation within a wider social context, but their accounts are still incomplete. The defenders of global regulation have no persuasive explanation of why states of the North can be trusted to promote sustainable use. The advocates of communal management, while they often acknowledge the importance of external factors in determining the success of devolved proprietorship, have not yet demonstrated that it is always possible to realize a supportive context for local management. If the 25 years of CITES has taught us anything, it should be that conservation policy is much more complex and involves many more different types of consideration than was originally assumed by CITES.

REFERENCES

Bonner, R (1993) *At The Hand of Man*, Simon and Schuster, London
Bromley, D (1991) *Environment and Economy*, Blackwell, Oxford
Carruthers, J (1989) 'Creating a National Park, 1910–1926', *Journal of Southern African Studies*, Vol 15, No 2, pp 188–217
Favre, D (1993) 'Debate within the CITES Community: What Direction for the Future?', *Natural Resources Journal*, Vol 33, pp 875–918
Grove, R (1995) *Green Imperialism*, Cambridge University Press, Cambridge
Hardin, G (1968) 'The Tragedy of the Commons', *Science*, Vol 162, pp 1243–8
MacKenzie, J (1988) *The Empire of Nature*, Manchester University Press, Manchester
Murphree, M (1994) 'The Role of Institutions in Community-based Conservation', in Western, D and Wright, R (eds) *Natural Connections*, Island Press, Washington DC
Ostrom, E (1990) *Governing the Commons*, Cambridge University Press, Cambridge
Ranger, T (1989) 'Whose Heritage? The Case of the Matobo National Park', *Journal of Southern African Studies*, Vol 15, No 2, pp 217–249
Swanson, T (1994) *The International Regulation of Extinction*, Macmillan, Basingstoke

Part V

ENDPIECE

Chapter 15

The Lesson from Mahenye

Marshall W Murphree

INTRODUCTION

During the eighth COP, in March 1992, I delivered an address to the Kyoto Forum, 'In Harmony with Wildlife', in which I tried to draw out the conjunctions linking communities in the wildlife-rich areas of rural Africa with CITES and the dissonance which lay within this linkage. To illustrate the view from below I used the experiences and perspectives of the people of Mahenye, a community in Zimbabwe's south–east lowveld that was struggling to put in place its own version of conservation through the sustainable use and management of its wild flora and fauna. Subsequently, this address was published in revised form under the title *The Lesson from Mahenye: Rural Poverty, Democracy and Wildlife Conservation* (Murphree, 1995). The editors have graciously invited me to provide a reflective endpiece to this volume, using the same theme and title. With the data from seven additional years of the Mahenye experience since the Kyoto address now available, this provides the opportunity to test my assertions that the people of Mahenye have the capacity and motivation to make responsible decisions about their natural resources and implement them with greater efficiency than any regulatory regime imposed from above. It also provides the opportunity to examine whether the decisions of CITES over these seven years have had any impact on what has happened in Mahenye and, if so, what one can infer about this linkage between global regulation and communal management.

THE EVOLUTION OF COMMUNITY MANAGEMENT

To understand what is now happening in Mahenye we must understand what has happened in the past. Mahenye is a community at the extreme southern end of the Chipinge District in Zimbabwe, located in a wedge of land (210 sq km) between the Mozambique border to the east and the Save River which forms its boundary with the Gonarezhou National Park to the south-west. Average rainfall is low (450–500 mm *per annum*), supporting the successful dry land cultivation of grains only in good seasons. Most of this area is covered by mixed mopane and combretum woodland, but along the Save River dense riverine forest occurs, supporting a broad range of floral and avian species, some of them rare in Zimbabwe. Ngwachumene Island, within the borders of the ward, is particularly noted in this regard (Booth, 1991). The alluvium soils in and around this riverine forest are used for the cultivation of vegetables and dry season grains.

The people of Mahenye are a segment of the Shangaan-speaking peoples who in pre-colonial times occupied the south-east lowlands of what is now Zimbabwe, the north-east lowlands of what is now South Africa and much of what is now the Gaza and Inhambane Provinces of Mozambique (see Junod, 1927 and Earthy, 1933 for early historical and ethnographical detail). In South Africa and Zimbabwe, colonialism brought displacement to these peoples and both the Kruger and Gonarezhou National Parks were carved out of Shangaan territory. In Zimbabwe the Shangaan were forced onto communal lands in the Chiredzi District and into Mahenye Ward in the Chipinge District. Some emigrated to Mozambique.

Thus, for the people of Mahenye the double expropriation of rights to land and wildlife brought by colonialism, and discussed earlier in this volume, was an acute reality. Evicted from their traditional territory and forced to live on its perimeter, they were also denied the right to use wildlife on either their old or their new land. This exclusion from the use of wildlife was for them not a minor denial. Hunting had been a major component in their livelihood strategies and their culture had evolved well-defined regulatory practices to make hunting off-takes sustainable (see Junod, 1927 and Wright, 1972).

It is not surprising, therefore, that at Zimbabwean independence in 1980 the people of Mahenye hoped to regain their homes in Gonarezhou National Park and have their hunting rights restored. In the event, the new government reasserted its claims to the park and park officials intensified their anti-poaching raids into Mahenye. In response to questions as to why the Park had not been returned to the Shangaan, government officials suggested that the government needed the foreign exchange brought in by international visitors to the Park. The people of Mahenye retorted that this gave them an added incentive to poach within the Park: if there were no animals there would be no tourists, if there were no tourists there would be no revenue and no reason for the government to retain the Park. Furthermore, elephant and buffalo from the Park raided their gardens and fields regularly, threatening the remaining means of subsistence left to them. As one elder put it to Peterson, 'We were displaced and we don't have the game we used to have. We have to depend on our agriculture. We must grow our crops to survive. Their animals cross over the river and eat our crops. We have no food when our crops are eaten, so we have to eat their animals or we would starve. If they would control their animals, we could grow our crops. Then there would be no poaching.' (Peterson, 1991, p 10).

The language as well as the substance of this quotation is significant. Wildlife, previously an important aspect of subsistence, had become 'their animals'. A spiralling dialect of confrontation had developed which resulted in the abrogation of the Shangaan ethic of sustainable hunting and an increasingly zealous, if ineffective, campaign of anti-poaching incursions by park staff into the community. Subsistence poaching by the people of Mahenye continued, both within their borders and in the Park. They also became indirectly involved in the commercial poaching of elephant, which was a serious problem during the late 1970s and early 1980s. Shadrack, a Shangaan hunter who led a team of poachers which took an estimated 20–30 bull elephants *per annum* for several years, evaded capture by moving between the Park, Mozambique and Mahenye, where the people turned a blind eye to his activities.

In effect, the state conservation agency, the Department of National Parks and Wildlife Management (DNPWLM), and the people of Mahenye had reached what Lee refers to as 'socially constructed stalemate' (Lee, 1993, p 12) in their respective conservation objectives through the exercise of effective vetoes that each possessed. By virtue of their in-place location and ability to defy externally imposed regulation, the Mahenye people were in a position to render enforcement by DNPWLM largely ineffective. By virtue of its statutory ability to deny the people of any rights to the use of wildlife on their land, DNPWLM precluded any incentive or space for them to develop their own regime of self-management and control in the use of this wildlife. It is this socially constructed stalemate, founded on the exercise of mutual vetoes, which lies at the heart of the disjunction in the linkages between national and global regulation and community management and to which this chapter will return in its concluding section.

In early 1982 the Mahenye leadership, through their councillor, arranged to have a meeting with the Gonarezhou warden in an effort to improve their relationship with the Park. Anticipating a difficult meeting, the warden, a white Zimbabwean, asked a local rancher and safari operator, Clive Stockil, to accompany him and provide translation. Stockil had learned Shangaan as a child, was well known and liked in the Mahenye community and respected in both local farming and government circles. The warden's anticipation of a difficult meeting was fulfilled as the Mahenye people put forward their case aggressively. Stockil changed his role from that of translator to instigative facilitator. With his grasp of the importance of wildlife to Shangaan culture and his personal experience of the devolution of wildlife proprietorship to private ranchers through the 1975 Parks and Wild Life Act, he seized on the importance of the pronoun in the complaints to park officials about 'your wildlife'. Was it possible, he speculated, to effect a similar change of proprietorship over wildlife to the Mahenye community? He put this to the meeting along the following lines: if government were to agree that wildlife within the ward would be regarded as the property of the community, with community rights to sell this wildlife on the safari hunting market and keep the proceeds, would Mahenye be willing to accept a degree of crop damage and cooperate in the repression of poaching?

The proposal was received with both interest and scepticism. It seemed like a step towards the re-appropriation of wildlife but there was doubt about whether the government would really implement such a suggestion Nevertheless, agreement was reached to proceed with negotiations. Subsequently, DNPWLM was approached and agreed to a one-year trial. Permits for two trophy elephant to be taken in Mahenye by clients found by Stockil would be issued for 1982. Provided that the community cooperated with park officials in anti-poaching efforts, DNPWLM would insist that the concession fees for the hunts would be paid to the Ward.

In August 1982 the hunts took place, with the people of Mahenye celebrating success in the ritually directed distribution of meat which followed. With a renewed hope of being given the authority to manage their natural resources, the community leadership persuaded those living on Ngwachumene Island (approximately 100 people in seven hamlets) to move to the mainland, thus leaving this prime habitat free for wildlife. At the same time, the level of poaching in the ward dropped dramatically. There remained, however, the matter of the concession fees promised by DNPWLM. Not for the last time, Mahenye was to learn that, in dealing with government, it had to deal with a number of its branches. The money had been received and lodged with the Treasury which insisted that it could only be returned via the Ministry of Local Government and the Chipinge District Council, which would have the right to determine its use. DNPWLM remained adamant that it had been promised to Mahenye. An impasse resulted and safari revenues for not only 1982 but also the following three years accumulated as the bureaucrats argued and procrastinated. In Mahenye suspicions increased that the whole affair was a scheme to clear Ngwachumene Island of settlement, deceive them into compliance and line a few private pockets.

By 1986 DNPWLM was in the final stages of preparing its proposals for CAMPFIRE and the Mahenye case was an embarrassment to it. It pressed the district council for a resolution of the matter, stating that $30,461[1] had accumulated in this account. During meetings held in October and November 1986, the district council debated the use of this money, with a majority of the councillors arguing that it had accrued from wildlife which should be regarded as the common property of the district and thus distributed throughout the district. The Mahenye representatives retorted that they regarded their wildlife as livestock, with the management and opportunity costs that livestock entails. If the district council wished to regard all livestock in the district as common property they would be content with the approach and assume that they would share in the proceeds of all livestock sales in the district in the future. This rationale had strength, but it was left to Chief Musikavanhu, from the northern end of the district, to clinch the argument. 'My children' he said, 'What we have heard is the truth. We have no claim on this money. We did not sleep in the fields to protect the crops from elephants, as the people from Mahenye did. The elephants are theirs, not ours' (see Stockil, 1987). The outcome of debates like this was that the district council agreed that the funds should go to Mahenye.

In February 1987, the $30,461 was presented to the community in the form of a cheque, to be used (as agreed by the Mahenye people) for a school building, a grinding mill and teachers' accommodation. Safari hunting continued in the ward during 1987–1990 under an agreement between Stockil, the Mahenye wildlife committee and the district council. However, once again, the proceeds were held up by bureaucracy, and it was not until January 1990, when the Chipinge District Council was granted appropriate authority status under the CAMPFIRE programme, that the way was clear for Mahenye once again to be in receipt of wildlife revenues generated prior to and after that time. Since 1990 Mahenye's safari hunting and wildlife management activities

---

[1]   Figures in this chapter are in Zimbabwe dollars.

have been carried out under arrangements generally reflecting the profile that CAMPFIRE exhibits elsewhere. A locally elected ward wildlife committee plans and carries out management functions, employing
local field staff to monitor wildlife, poaching and the hunting activities
of the professional hunter. The committee also sets budgets and is responsible to general community meetings for its activities and planning. The district council formally awards the hunting concession and
is in receipt of the concessionaire's payments, with the expectation
that these will be passed on to the Mahenye wildlife committee after
the deduction of levies and administrative fees.

COMMUNITY MANAGEMENT IN MAHENYE SINCE 1992

This was Mahenye's situation when my Kyoto address was prepared
in March 1992. They had created their own structures for the planning
and implementation of a wildlife utilization programme, had trained
and employed locally recruited wildlife monitors, had reached a point
where they were in receipt of wildlife revenues of $180,000 by 1992
and had developed a collective decision-making process for the determination of the allocation of these revenues.[2] These achievements had
been attained in the face of severe contextual constraints. Mahenye
had been accorded only a delegated and conditional proprietorship
over its wildlife resources, which in terms of Zimbabwe's current legislation, is vested in the district council. This means that revenue generated from its wildlife is subject to a 20 per cent administrative fee
levied by the council.[3] It also means that the community cannot negotiate directly with its clients in the market, such negotiations having
to follow a ponderous bureaucratic route which does not conform to
the time-frames required for successful entrepreneurship.

In spite of these constraints, Mahenye has significantly expanded
its success in community management since 1992. It has used the
limited authority and responsibility accorded to it as a stepping stone

---

[2]   These allocations fall into three categories ie, management, community
projects and direct dividend payments to the registered households of the
community. Statistics show annual fluctuations in allocations to these three
categories, but generally about 15 per cent of revenue goes to management,
35 per cent to community projects and 50 per cent to household dividends.

[3]   It is difficult to argue that the figures produced by this levy represent actual
and administrative expenses incurred by the council and they can more properly be considered as a council tax on the Mahenye enterprise.

to increment the scale of its natural resource management regime and augment the benefits flowing from sustainable exploitation. A number of factors have contributed to this, including the social cohesiveness and energy that the community is fortunate to possess. For the purposes of this analysis I will, however, focus on two other factors which have critically shaped this success: the innovative dynamic which has been unleashed by their qualified empowerment and the impact of their involvement with the private sector in the wildlife tourism industry.[4]

In Mahenye's case gaining proprietorship over wildlife resources provided the incentive for a broader set of land and resource-use planning initiatives. By the end of 1991, the community had developed a plan for their area which included the following:

- a safari camp at Chivirira Falls, to be used for both hunting safaris and game viewing visitors to Gonarezhou National Park;
- the development of an exclusive game viewing wildlife area in the north of the area, to include fencing, water points and the re-introduction of certain species which had been present in the past but which were now rare in the area;
- the development of a small scale irrigation scheme below Chivirira Falls; and
- the development of a crocodile and ostrich farm on Gombe Island.

This scheme was developed by Mahenye and not by any NGO or development agency, although they did consult Agritex, the government's agricultural extension agency, for technical advice. The integration of wildlife and agriculture is worth noting. The irrigation scheme, properly fenced, was considered compatible with wildlife uses and a rational use of the water available to them in the Save River. They wished to venture into game farming, using crocodiles and ostriches. And, in planning for a game viewing area within their area, they sought to reduce their dependency on access to Gonarezhou National Park as the basis for attracting tourists. Examples of successful game tourism and game farming on private ranches in Zimbabwe were known to them; they sought to draw on the experience, adapt it to their own circumstances and in so doing augment the self-sufficiency and autonomy of their growing wildlife and tourism enterprise.

---

4   For a much more detailed analysis of Mahenye's involvement in private sector tourism see M Murphree (forthcoming).

This scheme still stands as a general map for the community's development planning, but progress on its components has been varied. The irrigation scheme (for which the community is most dependent on government approval and support) has been approved by government but still awaits implementation. The crocodile and ostrich farm is still in abeyance and is unlikely to develop unless the current depression in markets for its products is overcome. It is in the plan's eco-tourism components that most progress has been made, which brings us to the topic of Mahenye's involvement with the private sector eco-tourism industry.

In 1993 the community entered into a partnership with Zimbabwe Sun Ltd (ZSL) for tourism lodge development in Mahenye. ZSL is a large hotel and tourism corporation with multinational operations. It wished to add Mahenye to its suite of regional tourist destinations, particularly as a gateway for its clients' entry into Gonarezhou National Park and as a site for niche tourism with emphasis on the floral and avian rarities found in the community's riverine forest. This partnership required the concurrence of the district council, which was forthcoming on the condition that a formal memorandum of agreement would be developed embodying the understandings of the partners. On the basis of this contractually tenuous arrangement, ZSL commenced the construction of two tourist lodges, the Mahenye Safari Lodge, next to Ngwachumene Island, and Chilo Lodge, approximately three kilometres upstream, overlooking the Chilo Gorge. Smaller and more intimate, Mahenye Lodge accommodates 16 guests and was built in 1993, opening in February 1994. Chilo Lodge, larger and more luxurious, accommodates 28 guests. Its construction was completed in September 1996.

While these developments were taking place the drafting of the formal memorandum of agreement was taking its ponderous course and was not signed (by the council and ZSL) until late 1996. The memorandum is a ten-year lease agreement requiring Mahenye to ensure that the lodge sites are 'rendered free from any disturbance arising either from human settlement or livestock'. ZSL is required to pay to the council annual payments of 8 per cent of its gross trading revenues in the first three years, 10 per cent in the next three and 12 per cent in the remaining four years. It also stipulates that ZSL will employ local labour 'wherever possible'. Council is committed to deduct no more than 20 per cent of the revenues paid to it by ZSL, with the balance going to the Mahenye community, 'it being understood by the parties that the payment of such revenues is of fundamental importance to this agreement'. Collectively, the memorandum binds the parties to:

> '*jointly endeavour to administer the Unit so as to ensure
> that the activities of ZSL are rendered as efficient and as
> profitable as possible and so as to ensure the proper and
> efficient preservation, management, responsible usage and
> protection of the natural habitat and wildlife found in the
> Unit, and further to ensure that the usage by ZSL is as
> unobtrusive and beneficial to the Mahenye Ward Commu-
> nity as possible*'[5].

ZSL estimates the costs of the construction and equipping of Chilo
Gorge Lodge at $16m and Mahenye Safari Lodge at $7m. In addition,
a 45km electricity supply line from the national grid at Quinton Bridge
had to be built, the gravel road from Quinton Bridge to Mahenye had
to be improved and a water purification, pumping and reticulation
system had to be installed. An airstrip was built near the lodges and a
telephone line was also put in place. These are estimated to have added
$1.9m to the capital start up costs, totalling $24.9m. With a capital
investment of this magnitude, it is clear that for ZSL the Mahenye
venture is a long-term investment. The investment is being made as a
business gamble on the future of the up-market Zimbabwean tourism industry which is currently performing well. Moreover, as an international company in an industry where environmental image is
important, ZSL's involvement in a community-based natural resource
management enterprise enhances its image of being an environmen-
tally responsible business. The company has a mission statement that
asserts its commitment 'to participate responsibly in the controlled
social and environmental development' of the regions where it operates and its Mahenye operation is seen as an example of this. Finally,
there is an element of political investment in the involvement. Within
Zimbabwe CAMPFIRE is popularly seen as a programme for rural,
black development and any large multinational which associates itself with this does its political image no harm.

The community's assessment of the tourism lodge development is
generally highly positive. Interestingly, community infrastructural development related to the lodge enterprise is usually mentioned
first in any listing of benefits. Among the benefits cited is improved
road transport. The improvement in the 45km road from Mahenye
to Quinton Bridge is considered a great asset, something that the

---

5   All quotes are from the 'Memorandum of Lease Agreement for communal land for
      trading or other purposes made and entered into between the Chipinge Rural
      District Council and Zimbabwe Sun Limited', Chipinge Rural District Council
      files, 1996.

community had been asking for from the council for many years without success. The advent of mains electricity is also mentioned. When the lodge development brought mains electricity to the lodge sites, the community seized the opportunity to arrange a further extension to the main centre of the settlement. This was at a cost of $140,500, which was advanced to the community against annual payments, the first repayment instalment to be $30,239. Currently electric connections are to the clinic, the school, the grinding mill and the police post. One trader has also connected his store to the mains and it is anticipated that other private users will have electricity installed. All supply is metered and charged to the respective users. The community takes great pride in having electricity in their remote location. Once again the sentiment is that government would never have got around to assist them with the electricity connection and that it is there because of their enterprise. One Wildlife Committee member proudly commented to me that the project had 'lit up this community in two ways... it has lit up our eyes and it has lit up our minds'. Finally, the community is pleased with the water reticulation that the lodges have brought. One of the conditions of the lease was that livestock would be excluded from areas immediately adjacent to the lodges. This was a provision with conflict potential, and to reduce this possibility ZSL agreed to provide livestock watering points away from the river. The community then requested that this reticulation be extended to the village centre. An agreement was reached whereby ZSL provided the water pipe line and supply, and the community dug the required trench.

While the people of Mahenye place great emphasis on the infrastructural benefits of tourist lodge development, the revenues of their eco-tourism enterprise remain a critical factor in their collective assessment. Table 15.1 shows how the inclusion of lodge tourism in their natural resource enterprise has influenced the level of these revenues.

Table 15.1 *Mahenye Tourism Revenues by Category, 1991–1997*

Year	Hunting safari revenue	Tourism lodge revenue	Total
1991	68,800	–	68,800
1992	180,000	–	180,000
1993	158,000	–	158,000
1994	163,736	–	163,736
1995	138,445	5,940	144,445
1996	138,495	140,484	278,979
1997	188,740	429,804	618,594

Source: Data from Chipinge Rural District Council and Mahenye Wildlife Committee files

Two salient points arise from this data. Firstly, we can note that by 1992, hunting safari revenues had reached a plateau which has remained in the range of $140,000–$180,000 per annum since then.[6] The revenue generating potential of the Mahenye hunting concession is almost entirely dependent on trophy elephant. Other safari species, such as leopard, buffalo, bushbuck, grysbok and impala, are present, but in low numbers. Elephant in Mahenye are part of a larger population that ranges across Gonarezhou National Park and Mozambique, and sustainability considerations limit Mahenye's quota to 2–4 trophy elephant per year. A rise in the price of elephant trophy prices could increase these revenues in the future, but it is clear that the capacity of the ward to expand its safari hunting revenues is finite and unlikely to grow significantly in the future.

Secondly, we can note that by entering the tourism lodge industry Mahenye broke through this off-take determined threshold of revenues. After a slow start-up period when the lodges were being built, lodge revenues overtook hunting revenues in 1996 and far exceeded these in 1997, representing 70 per cent of total revenues in that year. To the figure of $618,514 generated by lease fees can be added the wage receipts of Mahenye employees at the lodges, which in the same year totalled $413,356.

In summary, by 1997 the people of Mahenye were generating income of over a million dollars per year from their natural resources. These resources had also provided them with an improved road, mains electricity and water reticulation, and they had achieved these things without any direct donor or government aid. The community had exploited the limited space given them by the state to use their own resources and had linked their entrepreneurial energies to the enlightened self-interest of the private sector. Mahenye is no longer a community of subsistence farmers, where wildlife is a liability rather than an asset. It now has a collective natural resource enterprise, which forms an important component of the livelihood and investment strategies of its members. The incentives to guard their natural capital have been internalized. Ineffective, externally imposed regulation has been superseded by the imperative of sustainability in the use of this capital.

---

6 When inflation is considered, these figures indicate a revenue decline in real terms.

MAHENYE AND CITES

How have the decisions of CITES impacted on these developments in Mahenye? Directly, very little, if at all. CITES, the exemplar of global regulation used in many chapters of this book, is formally a treaty on international trade in endangered species. None of the species listed by CITES as 'endangered' are used for international trade in Mahenye's natural resource enterprise. The safari hunting of elephants was the economic springboard for Mahenye's enterprise and is still a component in this enterprise, but although the elephant is listed on Appendix I of CITES, this mode of use is not defined by CITES as 'commercial trade'. At the tenth COP held in June 1997, CITES agreed to downlist Zimbabwe's elephant population to Appendix II and to a trade between Zimbabwe and Japan in ivory under specified conditions. While financially important for Zimbabwe's state conservation budget, such a trade will have little direct financial impact on Mahenye's wildlife enterprise. They have no ivory to sell, except for a few tusks credited to the Chipinge Rural District Council's account held in DNPWLM's ivory stores in Harare and recovered from past poaching and natural mortality. Any revenue from such sales would have only a marginal effect on their annual budgets and does not enter into their forward planning.

We must note, of course, that had Mahenye's plans for the development of an ostrich and crocodile farm been implemented, CITES regulations would have intruded more directly on their management, increasing transaction costs as they dealt with bureaucratic procedures and paperwork. We must also note that Mahenye's situation does not reflect that of other communities in Zimbabwe's CAMPFIRE Programme, where an international trade outlet for ivory, elephant products and live sales could make a difference in their revenues.[7] Nor does it reflect the situation of communities elsewhere, where international trade in species listed by CITES forms a central aspect of collective enterprise and household livelihoods. Staying with our case study, however, we must conclude that the operations of CITES and Mahenye's natural resource management show little in the way of direct causal relationship. There is, however, a relationship between the two. This relationship is indirect, mediated and diffuse, but critically important nevertheless.

---

[7]   One estimate of the value of communal land ivory stocks held in Zimbabwe's central ivory repository suggests that these are worth in excess of US$3 million (Child, 1995).

Firstly, we can note that the decisions of CITES can have an influence on the policies of its member states regarding the locus of efficient regulation. CITES places the onus of regulation on the state, assuming that its members have the will and capacity to enforce regulation but recognizing that this may not necessarily be the case in its 'non-compliance' provisions. State agencies, operating in the same bureaucratic culture, thus tend to be compliance-driven in their policies of regulation. There are exceptions, such as Namibia and Zimbabwe. In these two countries, policy has evolved over the past 20 years towards a prioritization of efficiency in regulation with a concomitant emphasis on the devolution of regulatory responsibility for wildlife outside state lands to private or communal land holders.[8]

Zimbabwe, in its case for the downlisting of its elephant population to Appendix II at the tenth COP, used the success of this devolutionary approach to regulation as one of its arguments. This was done in spite of the fact that this devolution, in the case of communal lands, was attenuated and incomplete, as the Mahenye example illustrates. Having won its case, we can now expect pressures to exert themselves on Zimbabwe's regulatory policy stance at two levels. At the bureaucratic level, a compliance orientation will be emphasized to ensure that Zimbabwe satisfies the conditionalities attached to its entry into the ivory trade. At the political level, pressure will rise for Zimbabwe to fully implement in communal contexts the devolutionary approach to regulation that it has proclaimed as policy. There are signs that Government is sensitive to this pressure. In a speech given seven months after the tenth COP, the Minister of Mines, Environment and Tourism announced that the Government was 'embarking on a process of continually easing legislation governing access by rural communities to all natural resources'. Observing that 'Over the years communities have attained the capacity to run financially stable ventures, to count game, to market their products, to prioritize their needs and make national decisions,' he went on to state that 'we envision a situation where communities have the legislative backup when they decide to go into business transactions and to negotiate deals on their own'.[9] Whether this legislation will be forthcoming remains to be seen but, if it is, a CITES decision (ostensibly on elephants) may well have provided – through its political dimensions – an added impetus to the provision of the legal status which Mahenye needs for its

---

8   For an extended discussion see B Jones and M Murphree (forthcoming).
9   Speech by the Honourable Mr S K Moyo, Minister of Mines, Environment and Tourism, delivered 26 February 1998 and reproduced in *CAMPFIRE News*, No 17, March 1998.

own programme of regulation and use. It is in such chains of indirect, mediated causality that the interests of community management and the actions of CITES can be linked positively.

There is, however, an aspect of this linkage that can frustrate its potential for synergy. The chain links together two polarities that march to different drummers and that operate within different regulatory cultures. CITES is historically rooted in the industrialized and urbanized societies of the developed world, with their emphasis on the existence and recreational value of wild places and wild species. Its implementation is largely in the hands of an international epistemic community of scientists, bureaucrats and agency professionals who place great faith in the predictive capacity of science and the efficacy of legal proscription. It tends therefore to be reductionist and compliance-oriented in its thinking and in its approach to regulation.

Community management, as exemplified by our Mahenye case study, is faced with a different set of incentives and regulatory dynamics. It values wild flora and fauna instrumentally, as a means to human livelihoods, and the incentive for conservation lies in the desire to maintain this natural capital. It is inductive and experimental, accepting contingency and risk. It is adaptive and opportunistic, aligning itself more closely with the opportunities of the market than with the promises and demands of the state. It is efficiency oriented, and regulation has its source in the need for communal compliance if the collective enterprise is to succeed.

Each of these polarities of regulatory culture has a role to play in the twenty-first century. The elimination of either is not a viable option in a world that requires the efficient achievement of multiple objectives at different levels. It should thus be a matter of profound concern that each polarity can adversely affect the efficiency of the other. Community management, through its *de facto* power to sidestep externally imposed regulation and its dynamic new alliances with the market, may make CITES irrelevant in practice.

CITES may, on the other hand, take decisions which, in their impact, have the potential to go beyond its mandate and push policies in directions which inhibit the conditions for the development of robust community regimes of regulation and use. Mahenye's story is again a case in point. As already mentioned, Mahenye's collective enterprise has never used its elephants for purposes of international trade, since its safari-hunted ivory trophies are excluded from CITES' definition of commercial trade. However, the CITES Appendix I listing of the elephant has been used by the anti-hunting lobby in the US in their recent attempt to persuade the US Government to ban the importation of safari-hunted ivory into the US under the provisions of the US Endangered Species Act and the CITES provision for 'stricter domestic measures'. Had this attempt been successful, Mahenye's enterprise would have been significantly affected, since its safari clients are largely from the US.

It is in such articulated and stochastic chains of causality that Mahenye and CITES are linked. In the search for regulatory efficiency these linkages can create an impasse, since both global regulation and community management have the capacity to veto each other. To convert impasse to synergy must be the strategic goal for global regulatory bodies such as CITES, and the scope of strategic thinking required must move well beyond that which CITES has evidenced to date. Such thinking must take into consideration the larger canvas of contemporary evolution in societal governance. In the words of Rock, the state:

> '*is being transformed and superseded in ways that are only tentatively beginning to be understood... Power is being dispersed and converted as new systems of knowledge, accountability and regulation replace more familiar methods of direct, centralized bureaucratic control. Society itself is becoming cast as a market in which people, once defined as citizens, are now customers'* (Rock, 1998, p 9).

These words were written against the background of changes in the post-industrial, urbanized societies of the North. However the analysis has a resonance with what has transpired in Mahenye. If CITES is to endure as more than a bureaucratic shell, it must incorporate these trends in its strategic planning. Importantly, it must give far more attention to the operations of the market. Swanson's prescription of external consumer market regulation (see, for example, Chapter 12) is an analytic step in this direction.

However, I share with Dickson (Chapter 14) his reservations about Swanson's assumptions concerning the role and capacities of producer and consumer states. Mahenye's case underlines the growing importance of local entrepreneurship and its links with the private sector. This is a trend that is consistent with the new systems of knowledge, accountability and regulation that are replacing older methods of centralized, bureaucratic control. It is a trend that CITES cannot ignore if it is concerned with effective conservation rather than mere institutional self-perpetuation.

In other words, CITES must reform its regulatory culture, aligning itself more closely with current societal trends in governance, accountability and incentive. It must reconstitute its constituency, giving priority to those who are the effective producers of wildlife and guardians of biodiversity. If it can do so it will create a new synergy between the local and global dimensions of conservation. If it fails to do so it will perpetuate the socially constructed stalemate which characterizes the current disjunction between global regulation and community management.

REFERENCES

Booth, V (1991) *An Ecological Resource Survey of Mahenye Ward, Ndowuyu Communal Land, Chipinge District*, WWF Multiple-species Animal Production Systems Project, Project Paper No 20, Harare

Child, B (1995) 'Can Devolved Management Conserve and Develop the Management of Natural Resources in Marginal Rural Economies?', unpublished paper, Department of National Parks and Wild Life Management, Harare

Earthy, D (1933) *Valenge Women*, Frank Cass, London

Jones, B and Murphree, M (forthcoming) 'The Evolution of Policy on Community Conservation in Namibia and Zimbabwe', in Hulme, D and Murphree, M (eds) *African Wildlife and African Livelihoods: The Promise and Performance of Community Conservation*, James Currey, Oxford

Junod, H (1927) *The Life of a South African Tribe*, Macmillan, London

Lee, K N (1993) *Compass and Gyroscope: Integrating Science and Politics for the Environment*, Island Press, Washington DC

Murphree, M (1995) *The Lesson from Mahenye: Rural Poverty, Democracy and Wildlife Conservation*, IIED, Wildlife and Development Series, No 1, London

Murphree, M W (forthcoming) 'Community, Council and Client: A Case Study in Ecotourism Development from Mahenye, Zimbabwe', in Hulme, D and Murphree, M (eds) *African Wildlife and African Livelihoods: the Promise and Performance of Community Conservation*, James Currey, Oxford

Peterson, J (1991) *A Proto-CAMPFIRE Initiative in Mahenye Ward, Chipinge District*, Centre for Applied Social Sciences (CASS), University of Zimbabwe, Harare

Rock, P (1998) 'Sociology Revisited', *LSE Magazine*, Vol 10, No 2

Stockil, C (1987), 'Ngwachumene Island: The Mahenye Project', *The Hartebeeste* (Magazine of the Lowvelt Natural History Society), No 19, pp 7–11

Wright, A (1972) *Valley of the Ironwoods*, Cape and Transvaal Printers, Cape Town

Index

Also available from Earthscan

THE COMMERCIAL USE OF BIODIVERSITY

Access to Genetic Resources and Benefit-Sharing

Kerry ten Kate and Sarah A Laird

'This is book is an invaluable resource and guide for further research. It is as practical as it is informative and should be in the hands of every policy-maker, entrepreneur and student interested in the state and fate of the world's living resources' **Professor Calestous Juma**, former Executive Secretary of the Convention on Biological Diversity

Hb £50.00 ISBN 1 85383 334 7

INTELLECTUAL PROPERTY RIGHTS, TRADE AND BIODIVERSITY

Seeds and Plant Varieties

Graham Dutfield

'This important book provides a detailed picture of the legal and scientific issues surrounding the international debate over how to protect genetic resources. Its meticulously researched and argued analysis should be read by all those involved in the debate as well as those with an interest in protecting access to, and research into, our natural heritage' **Margaret Llewelyn**, Deputy Director, Sheffield Institute of Biotechnology Law and Ethics, University of Sheffield

Hb £35.00 ISBN 1 85383 692 3

THE GREEN WEB

A Union for World Conservation

Martin Holdgate

Founded in 1948, IUCN is in truth a 'green web', a network linking nearly a thousand members, in most of the countries of the world. Those members include governments, national conservation agencies, and non-governmental bodies ranging from the highly scientific to the stridently activist.
Martin Holdgate, Director-General of IUCN between 1988 and 1994, gives us a fascinating history of half a century of work for conservation throughout the world.

Pb £17.50 ISBN 1 85383 595 1

PARTNERSHIPS FOR PROTECTION

New Strategies for Planning and Management of Protected Areas

edited by Sue Stolton and Nigel Dudley

With contributions from over 40 of the world's leading conservation experts, and resulting from a collaborative project between IUCN and WWF, the book sets out ways to safeguard all the major ecosystems and explores innovative management partnerships involving individuals, communities, companies and governments. It is essential reading and a vital tool for all those involved with or studying biodiversity and conservation.

Pb £18.95 ISBN 1 85383 609 5
Hb £45.00 ISBN 1 85383 614 1

Earthscan Publications Ltd, London
www.earthscan.co.uk